AF737887

UNIVERSAL DESIGN – BEST PRACTICE
VOLUME 1

Herausgegeben von Peter Zec
Edited by Peter Zec

reddot edition

INHALT CONTENT

Gutes Design steht für Lebensqualität. Gerade wer mit Behinderungen leben muss oder durch Alter oder Krankheit eingeschränkt ist, braucht optimal gestaltete Hilfsmittel, die ein weitgehend selbstbestimmtes Leben möglich machen und die Integration in Beruf und Gesellschaft erleichtern. Dass sich die Hilfsmittelindustrie dieser Herausforderung bewusst ist, hat der REHACARE DESIGN AWARD bewiesen. 67 internationale Aussteller der REHACARE INTERNATIONAL 2008 in Düsseldorf nahmen mit 96 Einsendungen an diesem Designwettbewerb teil und nutzten die Chance, sich mit ihren Produkten vor einer hochkarätigen Jury zu profilieren. Ein Ergebnis, auf das wir stolz sind und das Signalwirkung haben dürfte. Der Wettbewerb in Verbindung mit einer Sonderschau aller prämierten Exponate machte den Messebesuchern deutlich: Gutes Design ist in der Hilfsmittelindustrie ein Thema, das ernst genommen wird und Zukunft hat – und die REHACARE INTERNATIONAL Düsseldorf spielt dabei eine Vorreiterrolle. Ich freue mich, dass mit dem vor Ihnen liegenden Buch nun eine noch breitere Öffentlichkeit Gelegenheit hat, sich über designorientierte Innovationen aus den Bereichen Rehabilitation und Pflege zu informieren. Viel Spaß bei der Lektüre!

Good design spells quality of life. Notably people living with disabilities or hampered by old age and infirmity need optimally designed assistive devices to help them lead largely independent lives as well as to further their integration at work and in society at large. The REHACARE DESIGN AWARD impressively bears testimony to how the assistive devices industry has internalised and responded to this challenge. Sixty-seven international companies who exhibited at REHACARE INTERNATIONAL 2008 in Düsseldorf submitted 96 entries to the competition, taking advantage of the opportunity to raise their profile by winning acclaim for their products from a high-calibre jury. We are very gratified by this response and feel confident that it makes an important statement. Combined with a special presentation featuring all the winning entries, the competition impressed upon visitors to the fair that good design is a topic that's here to stay and will continue to be taken seriously by the assistive devices industry – with REHACARE INTERNATIONAL in Düsseldorf playing a leading role. I am delighted that, through the book now in your hands, innovations focusing on the design aspects of tools for rehabilitation and care can now reach an even broader public. Happy reading!

Wilhelm Niedergöker
Geschäftsführer Messe Düsseldorf
Managing Director of Messe Düsseldorf

PETER ZEC

VORWORT FOREWORD

Design im medizinischen Bereich, und vor allem in der Rehabilitation, ist heute so wichtig wie nie zuvor. Angesichts des demografischen Wandels wird unsere Gesellschaft immer älter, der Hype um Jugend, Fitness und Gesundheit zugleich immer ausgeprägter. Hier entstehen neue Märkte, gilt es doch, die Rehabilitation für alle Generationen so integrativ wie möglich zu gestalten. Die Designer stehen angesichts des technologischen Fortschritts vor komplexen gestalterischen Herausforderungen. Nicht nur eine genaue Kenntnis der sozialen Bedürfnisse, sondern auch ein umfassendes Wissen der modernsten technischen Möglichkeiten wird von den Designern erwartet.

Vom Universal Design ist hier die Rede, einer umfassenden Gestaltung, die nicht nur das Wissen verschiedener Disziplinen vereint, sondern gleichzeitig die Bedürfnisse der unterschiedlichsten Zielgruppen zu gleichen Teilen berücksichtigt. Universal Design – das ist eine Erleichterung des Alltags für alle Altersgruppen, das sind intelligente Designlösungen, die einer klaren, modernen und hoch ästhetischen Formensprache folgen. Dabei gilt es, die technische Komplexität der Geräte und Hilfsmittel formal so stark zu vereinfachen, dass sie weitestgehend selbsterklärend sind – die wahrscheinlich größte Herausforderung, der sich ein Designer stellen kann.

Die Branche birgt ein enormes Potenzial. Es gilt, heute bereits die gestalterischen Grundlagen für das Design von morgen zu legen. Generationenübergreifend und ästhetisch einwandfrei. Der REHACARE DESIGN AWARD 2008/2009 zeigt, dass das Design in dieser Branche auf dem besten Wege ist. Und lässt uns gespannt in die Zukunft blicken.

Design in the medical field – and above all in the field of rehabilitation – has never been more important than today. In the face of demographic change our society is constantly growing older and the hype surrounding youth, fitness and health is increasing more and more. This is where new markets are developing, since the task is to make rehabilitation as integrative as possible for all generations. In view of technological progress designers are faced with complex design challenges. They are expected not only to have exact information regarding the social needs, but also comprehensive knowledge of the latest technical possibilities.

What I am referring to here is Universal Design, a comprehensive design which not only combines the knowledge of different disciplines but also takes into account the needs of different target groups in equal shares. Universal Design – that means making the daily life of all age groups easier, and intelligent design solutions that follow a clear, modern and highly aesthetic language of form. In this process it is necessary to formally simplify the technical complexity of devices and aids strongly, so that they are mostly self-explanatory – this is probably the biggest challenge designers can face.

The industry has enormous potential. The task at hand is to lay the foundations for tomorrow's design today; the design needs to serve all generations and be aesthetically impeccable. The REHACARE DESIGN AWARD 2008/2009 shows that design is well on the way in this industry, and causes us to look expectantly into the future.

Prof. Dr. Peter Zec
Initiator des red dot design award, Essen und Senator des
International Council of Societies of Industrial Design (Icsid), Montreal
Initiator of the red dot design award, Essen and senator of the
International Council of Societies of Industrial Design (Icsid), Montreal

UNIVERSAL DESIGN
BETRACHTUNG AUS VERSCHIEDENEN PERSPEKTIVEN LOOKING FROM VARIOUS PERSPECTIVES

Universal Design – Produkt- und Lebensqualität als neues Paradigma unserer Zeit
Universal Design – Quality of product and life as a new paradigm of our time

Eine Einführung von Burkhard Jacob
An introduction by Burkhard Jacob

„Universal Design" gilt als Trend der Zukunft. Kleiner, schneller und intelligenter war gestern. Heute heißt es sicherer, verständlicher und benutzerfreundlicher. Die Produkte der Zukunft sollen für alle leicht zu bedienen sein: Jung und Alt, Mann und Frau. Alters- und Geschlechtsunterschiede sollen verschwinden, Behinderte sind gleichgestellt. Universal Design räumt auf mit überholten Ansichten und liefert neue Einsichten aus der Perspektive des Designs. Ethische Fragen werden dabei ästhetisch beantwortet und sensibilisieren dafür, dass der gesellschaftliche Umbruch die Chance zu einem gestalterischen Aufbruch bietet. Was aber genau ist Universal Design? „Universal Design" ist kein Schlagwort. Nicht im Design. Nicht in der Architektur. Nicht in der Wirtschaft. Und sicher nicht in der medizinischen Pflege und Rehabilitation. Universal Design ist ein Ausdruck dafür, dass es in unserer Gesellschaft nach wie vor Integrationsprobleme von Behinderten oder Älteren gibt. Eine Situation, die unser Wissen und Gewissen fordert.

Experte in eigener Sache

Ursprünglich geht der Begriff auf den US-amerikanischen Architekten Ronald L. Mace zurück, der schon als kleiner Junge sein gestalterisches Talent und seine technische Begabung entwickelte. Zusammen mit seinem Vater baute er Seifenkistenautos und mit seinem älteren Bruder nahm er an Wettbewerben für Modellflugzeuge teil. 1950, im Alter von nur 9 Jahren, erkrankte Ron Mace an Polio. Ein Impfstoff war noch nicht gefunden. Die Ärzte rieten seinen Eltern, ihn nicht zu Hause zu pflegen. Aber sowohl sein Vater als auch seine Mutter waren überzeugt, dass er in der häuslichen Umgebung und in der Familie am besten aufgehoben sei. Durch die Kinderlähmung wurde Ron Mace unweigerlich zum Experten in eigener Sache, immer auf der Suche nach kleinen Hilfen für den Alltag. Alles, was er brauchte, fertigte er selbst an – ganz gleich ob es ein geeignetes Bett war oder ein Rollstuhl, um ins Bad zu gelangen. „It's the set of

"Universal Design" is considered the trend of the future. Smaller, faster, and more intelligent was yesterday. Today's maxims are 'safer', 'easier to understand', and 'more user-friendly'. The products of the future are supposed to be easy to use for everyone, young and old, men and women alike. Age and gender differences are to disappear; people with disabilities are equal. Universal Design does away with outdated views and offers new insights from the perspective of design. In the process ethical questions are answered aesthetically and create awareness of the fact that the social change offers the opportunity for a new start in design. But what exactly is Universal Design? "Universal Design" is not a catchphrase, not in the field of design, architecture and economics, and certainly not in the area of medical care and rehabilitation. Universal Design is a term for the fact that there are still integration problems in our society with regards to handicapped people or the elderly. This is a situation that challenges our knowledge and our conscience alike.

Expert in personal concerns

The term was originally coined by U.S. architect Ronald L. Mace, who already as a young boy developed his design and technical talent. Together with his father he built soapbox cars and with his older brother he took part in model aircraft competitions. In 1950, when he was only nine years old, Ron Mace contracted polio. At that time a vaccination had not yet been discovered. The doctors advised his parents against having him stay at home and caring for him themselves. However, his father and mother were convinced that he would be best cared for at home and in the family. Polio involuntarily made Ron Mace an expert in his personal concerns, always searching for small aids for daily life. Everything he needed he built himself – no matter if it was a suitable bed or a wheelchair to get to the bathroom. "It's the set of the soul that determines the goal", he wrote into his high school yearbook, a quote from

the soul that determines the goal", schrieb er in sein High-School-Jahrbuch, ein Zitat der amerikanischen Schriftstellerin Ella Wheeler Wilcox („One Ship Sails East"), von dem er zutiefst überzeugt war. Nach seinem Abschluss wollte er Architekt werden und bewarb sich an der North Carolina State University, wo ihm der Dekan der „School of Design" davon abriet, das Studium aufzunehmen. Der Dekan glaubte, dass das Programm mit Rons Behinderung nicht zu bewältigen sei. An Studentenwohnheime, die man mit dem Rollstuhl hätte nutzen können, war nicht zu denken. Ron Mace merkte, dass die Welt, an der er teilhaben wollte, nicht so gestaltet war, um ihn wie jeden anderen Menschen zu integrieren. Zweifellos würden ihm die Räumlichkeiten der Universität Probleme bereiten, aber es schien ihm nicht unmöglich, ein erfolgreiches Studium zu absolvieren.

Also nahm er sein Studium auf. Seine Mutter unterstützte ihn. Sie mietete einen Wohnwagen an und zog mit ihm nach Raleigh. Da viele Räumlichkeiten der Universität für Rollstühle nicht zugänglich waren, arbeitete Ron die meiste Zeit des Studiums in seinem Wohnwagen. Und doch wurde er durch das, was er tat und wie er lebte, zum Vorbild für andere Menschen mit Handicap. Allein die Anwesenheit von Ronald Mace hatte die Einstellung vieler Professoren und Studienkollegen an der North Carolina State University verändert.

Um eine Zulassung als Architekt zu erhalten, brauchte er einen Praktikumsplatz bei einer Firma. Die Suche gestaltete sich aber schwierig. Und so machte Ron Mace erstmals mit der Erfahrung Bekanntschaft, aufgrund körperlicher Merkmale von der sozialen Welt ausgeschlossen zu sein. Als er im Kindesalter an Polio erkrankte, kämpfte er. Als er sich im Alltag und im Studium zurechtfinden musste, lernte er. Als er kurz vor seinem Ziel stand, merkte er, dass die Welt nicht für ihn gemacht war. Damals war ihm klar geworden, dass er seinen Wunsch, Architekt zu werden, nur in Abhängigkeit von anderen erreichen konnte. Er spürte die Ungleichheit am eigenen Leibe. Er konnte die Zähne zusammenbeißen, aber gegen Diskriminierung konnte er wenig tun.

Als Ron Mace 1966 mit einem Bachelor in Architektur graduierte, fand er schließlich doch noch als letzter seiner Studienkollegen einen Platz bei einer Firma in Greensboro, North Carolina, und arbeitete für einige Jahre als Architekt. 1972 bot ihm Charles Bell, ein befreundeter Studienkollege, eine Stelle als Dozent am Fayette Technical Community College an. Ron Mace nahm das Angebot an und arbeitete 1973 am ersten „Building Code for Accessibility" mit, der zum Vorbild für andere Bundesstaaten wurde. Er räumte auch mit der überkommenen Vorstellung auf, behindertengerechte Gebäude wären zu teuer, als er 1976 in dem Buch „Accessibility Modifications" die hohen Kosten als Mythos enttarnte. Seine Pionierarbeit war schließlich grundlegend für die nationale Gesetzgebung, die die Diskriminierung von versehrten oder behinderten Menschen verbot: Fair Housing Amendments Act (1988) und The Americans with Disabilities Act (1990). 1989 hatte Ronald Mace das mit öffentlichen Mitteln geförderte „Center of Accessible Housing" an der North Carolina State University ins Leben gerufen, heute bekannt als

American author Ella Wheeler Wilcox ("One Ship Sails East"), in which he deeply believed. After graduating from high school he wanted to become an architect and applied to North Carolina State University, where the dean of the "School of Design" advised him against studying. The dean believed that the programme was not manageable with Ron's disability. Student halls of residence which could be accessed by wheelchair were unimaginable at that time. Ron Mace recognised that the world he wanted to partake in was not designed to integrate him like any other person. Undoubtedly the premises of the university would cause him problems, but it did not seem impossible to him to complete a degree successfully.

So he started his degree. His mother supported him. She rented a caravan and moved to Raleigh with him. Since many premises of the university were not accessible by wheelchair, Ron studied in the caravan most of the time during his degree. He nevertheless became a role model for other people with disabilities. The presence of Ronald Mace alone has changed the attitude of many professors and fellow students at North Carolina University.

In order to get professional accreditation as an architect, he needed a work placement with a company. The search was very difficult, and Ron Mace for the first time experienced being excluded from the social world due to physical characteristics. When he fell ill with polio as a child, he fought. When he had to master daily life as well as the challenges of studying for a degree, he studied. When he was close to his goal, he became aware that the world was not made for him. Back then he realised that he could only fulfil his desire to become an architect depending on others. He experienced the inequality first hand. He could grit his teeth, but he could do very little against discrimination.

When Ron graduated with a Bachelor of Architecture degree in 1966 he eventually found a place with a company in Greensboro, North Carolina, as the last of his fellow students, where he worked as an architect for some years. In 1972 Charles Bell, a university friend, offered him a job as lecturer at Fayette Technical Community College. Ron Mace accepted the offer and in 1973 collaborated on the first "Building Code for Accessibility", which became a model for other U.S. states. He also did away with the obsolete notion that handicap accessible buildings were too expensive, when he exposed the high costs as a myth in the book "Accessibility Modifications" in 1976. Eventually his pioneering work became the foundation for national legislation prohibiting discrimination of disabled or handicapped people such as the Fair Housing Amendments Act (1988) and the Americans with Disabilities Act (1990). In 1989 Ronald Mace founded the "Center of Accessible Housing" at North Carolina State University, which was supported by public funds and is known today as the "Center of Universal Design". Here he was also professor of architecture at the School of Design, turning the institute into one of the leading research facilities in the field of Universal Design in the U.S.

„Center of Universal Design“. Hier war er auch Professor im Fachbereich Architektur an der School of Design und machte das Institut auf dem Gebiet des Universal Designs zu einer führenden Forschungseinrichtung der USA.

Feine Unterschiede, die entscheiden

Die ursprüngliche Idee des Universal Designs entwickelte sich also aus der Forderung nach Barrierefreiheit, einer Forderung, die von versehrten und behinderten Menschen selbst ausging und insbesondere die Unzugänglichkeit von Gebäuden und Räumen vor Augen führte. Architektonische Hindernisse zu entfernen bzw. barrierefreie Gebäude zu planen wird in den USA durch „Building Codes“ und die „ADA Standards for Accessible Design“ geregelt, die 1994 vom amerikanischen Justizministerium verabschiedet wurden. Gleichwohl gehen die Forderungen heute weiter, da die Forderung nach Barrierefreiheit nicht nur bauliche Aspekte, sondern auch Verkehrsmittel, Elektronik-, Informations- und Kommunikationssysteme sowie Dienstleistungs- und Serviceangebote betrifft. 2003 wurde daher in Europa die aktualisierte Fassung des „European Concept for Accessibility“ verabschiedet, die 2005 auch in deutscher Sprache vom Europäischen Institut Design für Alle in Deutschland e. V. mit Unterstützung gemeinnütziger Stiftungen und unterschiedlicher Vereine für Behinderte herausgegeben wurde.

Allerdings ist das Prinzip der Barrierefreiheit meist nur auf die Gruppe der Menschen mit Mobilitäts- oder Aktivitätseinschränkungen bezogen, obwohl die Barrierefreiheit im Interesse aller Menschen ist. Ronald Mace war sich dieser Tatsache vollends bewusst. In seiner letzten Rede, die er 1998 in New York auf der Konferenz „Designing for the 21st Century: An International Conference on Universal Design“ hielt, führte er den Anwesenden daher vor Augen, dass Barrierefreiheit für gewöhnlich eine Frage der Zugänglichkeit ist, die sich in erster Linie an den Interessen von Behinderten orientiert. Universal Design orientiert sich dagegen ganz allgemein am Gebrauch. Es ist nicht zielgruppenorientiert, sondern betrifft alle Nutzer oder Konsumenten, so Ronald Mace:

„Barrierefreies Design ist, was wir früher als Problem der Zugänglichkeit bezeichnet haben. Es ist vornehmlich eine Bewegung, die sich an Behinderungen orientiert. Architektonische Barrieren durch Baugesetze und -vorschriften zu beseitigen ist barrierefreies Design … Universelles Design dagegen definiert den Benutzer grob. Es ist ein vom Verbrauchermarkt angetriebenes Thema. Es orientiert sich nicht spezifisch an Menschen mit Handicaps, sondern an allen Menschen … Nun, bei assistierender Technologie handelt es sich meiner Meinung nach tatsächlich um Geräte für den persönlichen Gebrauch – und zwar die Dinge, die sich am Einzelnen orientieren – Dinge, die eine Funktion, die beeinträchtigt ist, ausgleichen oder unterstützen. Viele unter uns tragen eine Brille, weil unsere Sicht beeinträchtigt ist. Unsere Brille ist daher ‚assistive technology‘ – assistierende Technologie. Wir könnten also zu Recht sagen, dass jeder, der eine Brille

Subtle differences that make all the difference

The original idea of Universal Design developed out of the demand for freedom from barriers; a demand which came from people with disabilities and which highlighted the inaccessibility of buildings and rooms in particular. In the U.S., removing architectural obstacles, i.e. planning accessible buildings, is regulated by building codes and the ADA Standards for Accessible Design, which were passed by the American Justice Department in 1994. Nonetheless the demands continue today, since the demand for freedom from barriers not only applies to architectural aspects but also means of transport as well as electronic, information and communication systems and services. Therefore in 2003 the updated version of the "European Concept for Accessibility" was passed, which was also published in German by the 'Europäisches Institut Design für Alle in Deutschland e. V.' (EDAD), with the support of charitable foundations and different associations for people with disabilities.

However, the principle of freedom from barriers often only refers to the group of people who are restricted in their mobility or activities, even though accessibility is in everyone's interest. Ronald Mace was completely aware of this fact. In his last speech, which he held at the conference "Designing for the 21st Century: An International Conference on Universal Design" in New York in 1998, he demonstrated to the audience that freedom from barriers is normally a question of accessibility, which primarily focuses on the interests of people with disabilities. Universal Design, however, focuses on use in general. It is not target-group oriented but is aimed at all users and consumers instead, according to Ronald Mace:

"Barrier-free design is what we used to call the issue of access. It is predominantly a disability focus movement. Removing architectural barriers through the building codes and regulations is barrier-free design … Universal design broadly defines the user. It's a consumer market driven issue. Its focus is not specifically on people with disabilities, but all people … Now, assistive technology to me is really personal use devices – those things focused on the individual – things that compensate or help one function affected by a disability. Many of you wear eyeglasses because you have limited sight. The assistive technology is your eyeglasses. We could legitimately say that everybody who wears eyeglasses has a disability. Another example of assistive technology is my wheelchair. I need it as an individual. It is not a consumer product. It's for me. It's an assistive technology device."

To put it simply: every human being is in some way like every other person, but at the same time also like no other. Therefore Universal Design focuses on the use of things which are important to people. "Assistive Technology" attempts to take into account the characteristics of the individual. The term "Assisted Living" is used synonymously today.

trägt, eine Behinderung hat. Ein weiteres Beispiel für assistierende Technologie ist mein Rollstuhl. Ich benötige ihn als Individuum. Er ist kein Konsumartikel. Er ist für mich als Person. Er ist ein technologisches Gerät, das mich unterstützt."

Vereinfacht lässt sich sagen: Jeder Mensch ist in gewisser Hinsicht wie alle anderen Menschen, aber gleichzeitig auch wie kein anderer Mensch. „Universal Design" konzentriert sich daher auf den Gebrauch von Dingen, die für alle Menschen von Bedeutung sind. „Assistive Technology" versucht, die Besonderheiten des Einzelnen zu berücksichtigen. Synonym dazu wird heute auch der Begriff „Assisted Living" benutzt.

Prinzipien, die sich an Details messen lassen müssen

Bereits ein Jahr zuvor hatte Ron Mace gemeinsam mit Forschern, Architekten und Designern sieben Prinzipien des Universal Design beschrieben, die 1997 am Center of Universal Design der North Carolina State University veröffentlicht wurden.

1. Gleichberechtigte Nutzbarkeit: Das Design ist nützlich und marktfähig für Menschen mit unterschiedlichen Fähigkeiten.

2. Flexibilität im Gebrauch: Das Design kommt einer großen Bandbreite an individuellen Präferenzen und Fähigkeiten entgegen.

3. Einfache und intuitive Benutzung: Der Gebrauch des Designs ist einfach zu verstehen, unabhängig von Erfahrung, Wissen, Sprachfähigkeit und momentanem Konzentrationsgrad des Benutzers.

4. Wahrnehmbare Information: Das Design kommuniziert notwendige Informationen effektiv, unabhängig von der Umgebung oder den Wahrnehmungsfähigkeiten des Benutzers.

5. Fehlertoleranz: Die Gestaltung minimiert Risiken und schädliche Konsequenzen versehentlicher oder unbeabsichtigter Handlungen.

6. Geringer physischer Aufwand: Die Gestaltung ermöglicht effiziente und komfortable Benutzung mit minimaler Ermüdung.

7. Größe und Platz für Erreichbarkeit und Zugänglichkeit: Eine angemessene Größe und ausreichend Platz in Bezug auf Erreichbarkeit, Handhabung und Benutzung unabhängig von der Körpergröße, -haltung oder Mobilität des Benutzers sind gewährleistet.

Die Prinzipien wurden mit Beispielen versehen, die deutlich machten, dass es sich um Richtlinien handelte, die eine Orientierung bieten sollten. Allen Beteiligten war klar, dass diese Prinzipien nicht mit technisch fortschrittlichem, materiell hochwertigem oder wirtschaftlich erfolgreichem Design gleichbedeutend sein mussten. Sie ergaben sich vielmehr aus dem alltäglichen Gebrauch. Vor diesem Hintergrund erschien das Konzept von einem allgemein gültigen Design

Principles that have to be measured in the details

One year earlier, Ron Mace had, together with researchers, designers and architects, described the seven principles of Universal Design, which have been published by the Center of Universal Design at the North Carolina State University in 1997.

1. Equitable use: The design is useful and marketable to people with diverse abilities.

2. Flexibility in use: The design accommodates a wide range of individual preferences and abilities.

3. Simple and intuitive use: The use of the design is easy to understand, regardless of the user's experience, knowledge, language skills, or current concentration level.

4. Perceptible information: The design communicates necessary information effectively to the user, regardless of ambient conditions or the user's sensory abilities.

5. Tolerance for error: The design minimises hazards and the adverse consequences of accidental or unintended actions.

6. Low physical effort: The design can be used efficiently and comfortably and with a minimum of fatigue.

7. Size and space for approach and use: Appropriate size and space is provided for approach, reach, manipulation, and use regardless of the user's body size, posture, or mobility.

The principles were illustrated with the help of examples that make it clear that they are guidelines which offer orientation. Everyone involved understood that these principles did not have to be synonymous with technically progressive, high-quality material or commercially successful design. Instead they resulted from daily use. Against this background the concept of a universally applicable design seemed to be more of a search for an ideal. Even if this search could not come to an end, the idea was nevertheless desirable.

Multiplicity of terms as a reason for popular misunderstandings

The principle had been recognised, but it was not known well enough at that stage. Elaine Ostroff was also one of the researchers of the group of 1997. In 1978 she had co-founded the non-profit organisation Adaptive Environment, today's Institute for Human Centered Design in Boston, Massachusetts. Together with Wolfgang Preiser, professor of architecture at the University of Cincinnati, Ohio, she published the "Universal Design Handbook" in 2001. They gathered texts from more than 75 authors from North America, Europe, Asia, and Australia, thus drawing attention to the global

eher eine Suche nach einem Idealbild zu sein. Auch wenn diese Suche nicht zu einem Ende kommen konnte, so war doch die Idee erstrebenswert.

Begriffsvielfalt als Ursache für populäre Missverständnisse

Das Prinzip war erkannt, aber noch war es nicht hinreichend bekannt. Unter den Forschern der Gruppe von 1997 befand sich auch Elaine Ostroff, die bereits 1978 die Non-profit-Organisation „Adaptive Environment" mitbegründete, das heutige „Institute for Human Centered Design" in Boston, Massachusetts. Zusammen mit Wolfgang Preiser, Architekturprofessor an der University of Cincinnati, Ohio, gab sie 2001 das „Universal Design Handbook" heraus. Sie versammelten mehr als 75 Autoren aus Nordamerika, Europa, Asien und Australien und machten damit schließlich auf die globale Bedeutung des Universal Designs aufmerksam. Gleichzeitig wurde aber auch ein sprachliches Problem deutlich. Es existierten die unterschiedlichsten Begriffe und Vorstellungen. Der Bogen reichte vom „Accessible Design" über „Lifespan Design" bis zum „Inclusive Design" oder „Design for All". Es mochte an dieser Begriffsvielfalt liegen, dass sich der Begriff „Universal Design" nur langsam durchsetzen konnte, wenngleich diese Ausdrücke heute weitestgehend synonym verwendet werden.

Der Durchbruch des Universal Designs gelang erst vor dem Hintergrund des demografischen Wandels in den Industrieländern. In Deutschland löste insbesondere Frank Schirrmachers Buch „Das Methusalem-Komplott" eine breite Diskussion über das Altern der Gesellschaft und den bevorstehenden Generationenkonflikt aus. In einem provozierenden Stil beschrieb der Herausgeber der Frankfurter Allgemeinen Zeitung den bevorstehenden Wandel der alternden Industriegesellschaften. Seit den 1960er Jahren sind die Geburtenraten rapide gesunken. Gleichzeitig altern die Menschen der Industrienationen nahezu unbemerkt. Mit der Prognose, dass die Überalterung in den nächsten Jahrzehnten unumkehrbar ist und im Jahr 2050 allein in China so viele über 65-Jährige leben werden wie heute auf der ganzen Welt, macht unmissverständlich die Anforderungen und Herausforderungen klar, die nicht nur auf die Bereiche Design und Medizin zukommen, sondern auch die Gesundheits- und sozialen Sicherungssysteme der Industrienationen betreffen.

Wendepunkte: Chance und Risiko zugleich

Universal Design taucht also buchstäblich an Übergängen und Wendepunkten auf: Übergänge wie Türen, Treppen oder Brücken, Wendepunkte im Leben eines einzelnen Menschen, Übergänge von jüngeren zu älteren Generationen, Wendepunkte in der gesellschaftlichen Entwicklung. Übergänge, die uns vor Herausforderungen stellen, Wendepunkte wie im Leben von Ronald Mace und Peter van Nooy.

importance of Universal Design. At the same time a terminological problem became apparent. There was a diversity of terms and notions ranging from "Accessible Design" and "Lifespan Design" to "Inclusive Design" and "Design for All". It might have been due to the multiplicity of these terms that the term "Universal Design" could only establish itself slowly, even though these terms are mostly used synonymously today.

Universal Design had its breakthrough only against the background of the demographic change in the industrial countries. In Germany, Frank Schirrmacher's book "Das Methusalem-Komplott" (The Methusalem Plot) initiated a wide discussion about the aging of our society and the impending conflict between the generations. In a provocative style the editor of the Frankfurter Allgemeine Zeitung newspaper described the impending change of the aging industrial societies. Since the 1960s birth rates have decreased rapidly. At the same time people in industrial nations are aging almost unnoticeably. The prognosis that the increase in the percentage of old people will become irreversible in the next decades and that in 2050 in China alone there will be as many over-65-year-olds as there are today in the whole world makes the requirements and challenges unmistakeably clear which lie ahead not only in the areas of design and medicine, but also affect the health systems and the social security systems of the industrial nations.

Turning points: chance and risk at the same time

Thus Universal Design literally appears at transitions and turning points: transitions such as doors, stairs, and bridges, turning points in the lives of individual people, transitions from younger to older generations, turning points in the development of society, transitions that present us with challenges, turning points such as those in the lives of Ronald Mace and Peter van Nooy.

The Dutch-born Peter van Nooy had a car crash in 1990 and has been paraplegic ever since. When he visited the Motability Roadshow in his wheelchair in 2006 he saw a trike, a three-wheeled motor and recreational vehicle manufactured by the company Martin Conquest from Manchester. This was what he had waited for all those years in order to return to the street. Even though he had driven quads in the meantime, he was missing the convenience and freedom of riding a motorbike: "I'm thrilled; this Conquest will make an enormous change to my lifestyle. Effortless access to the driving position means I can be aboard and away in just a few minutes without any assistance at all, compared to the thirty minutes it takes me to get into my car. Now, I'll be able to travel independently, going when and where I want, and leaving the vehicle when I feel like it. It gives me back something that was taken away the day the car hit me – it gives me back my freedom. I can't convey how important this is to me," said Peter van Nooy.

Der gebürtige Niederländer Peter van Nooy verunglückte 1990 bei einem Autounfall und ist seit dieser Zeit querschnittsgelähmt. Als er 2006 in seinem Rollstuhl die Motability Roadshow besuchte, sah er ein „Trike“, ein dreirädriges Motor- und Freizeitfahrzeug der Firma Martin Conquest aus Manchester. Das war es, worauf er all die Jahre gewartet hatte, um endlich wieder auf die Straße zurückzukehren. Obgleich er in der Zwischenzeit auch „Quads“ gefahren hatte, vermisste er den Komfort und die Freiheit, die ein Motorrad bot: „Ich bin völlig begeistert, das Conquest wird meinen Lebensstil enorm verändern. Mühelos hinter das Steuer zu gelangen bedeutet für mich, dass ich in nur wenigen Minuten an Bord und unterwegs bin ohne jegliche Unterstützung, verglichen mit den 30 Minuten, die ich normalerweise brauche, um in mein Auto zu kommen. Jetzt kann ich mich unabhängig fortbewegen, wohin und wann ich will, und das Fahrzeug verlassen, wenn es mir gefällt. Es gibt mir etwas zurück, was mir an dem Tag, an dem mich der Wagen erfasste, genommen wurde. Es gibt mir meine Freiheit zurück. Ich kann überhaupt nicht ausdrücken, wie wichtig das für mich ist", so Peter van Nooy.

Das High-Performance-Motorrad ermöglicht es Rollstuhlfahrern, über eine automatisierte Rampe in das Motorrad zu rollen. Die rückwärtige Verriegelung wird über eine Drucktaste geschlossen und der Rollstuhl verschwindet hinter der Verkleidung. Zu Recht wurde es wie viele andere nennenswerte Produkte, die in diesem Buch versammelt sind, mit dem REHACARE DESIGN AWARD ausgezeichnet, weil der Gebrauch dieser Sache den Betroffenen ein Stück Freiheit zurückgibt und die Möglichkeit bietet, in einer Art zu leben, die sich an ihren Fähigkeiten orientiert. Das ist ein humaner Umgang mit einem sensiblen Thema.

Ähnlich wie Ronald Mace wurde Peter van Nooy Experte in eigener Sache. Und so wie Peter van Nooy geht es vielen anderen Menschen. Es überrascht daher nicht, dass die REHACARE als internationale Fachmesse und Kongress für die Bereiche Rehabilitation, Pflege, Prävention und Integration fast zur Hälfte von Betroffenen oder deren Freunden und Angehörigen besucht wird. „Es sind Experten in eigener Sache, die nach individuellen Lösungen suchen“, so Friedel Rinn, Präsident der REHACARE INTERNATIONAL und seit 1993 Vorsitzender der Bundesarbeitsgemeinschaft Selbsthilfe. „Keine andere Messe kann dieses Bedürfnis so umfassend befriedigen wie die REHACARE.“ So einfach dies klingen mag, so schwierig ist die Frage nach der richtigen Gestaltung zu beantworten. Sie muss immer wieder aufs Neue gestellt und entschieden werden. Personen mit einem Handicap, versehrte und ältere Menschen, aber ebenso Kinder und Schwangere benötigen gut gestaltete Produkte. In diesem Sinne ist Lebensqualität ein Maßstab des Lebens, nicht die Summe von Produkten oder Dienstleistungen. Lebensqualität schließt die Freiheit der Wahl und die Solidarität mit anderen Menschen ein.

The high performance motorbike enables wheelchair drivers to roll into the motorbike via an automatic ramp. The rear locking mechanism is operated by a push-button and the wheelchair disappears behind the fairing. This bike and other noteworthy products that are gathered in this book have rightly received the REHACARE DESIGN AWARD because the use of these products returns a piece of freedom to their users and offers them the opportunity to live in a way that focuses on their capabilities. This is a humane way of dealing with a sensitive topic.

Similar to Ronald Mace, Peter van Nooy became an expert for personal reasons. And many other people are in a similar situation as Peter van Nooy. Therefore it does not come as a surprise that half of the visitors of the REHACARE as an international specialist trade fair and congress for the areas of rehabilitation, care, prevention and integration are people affected or their friends and family. "They are experts in their own concerns looking for individual solutions," says Friedel Rinn, President of the REHACARE INTERNATIONAL and chairman of the 'Bundesarbeitsgemeinschaft Selbsthilfe', the association of self-help associations for people with disabilities and chronically ill people and their relatives in Germany, since 1993. "No other trade fair can take into account this need as comprehensively as the REHACARE." As simple as that may sound, as difficult it is to answer the question as to what is the right design. It has to be asked and decided anew every time. People with disabilities and the elderly, but also children and pregnant women need well designed products. In this sense, quality of life is not the sum of products or services but a benchmark of life. Quality of life includes the freedom of choice as well as solidarity with other people.

Ästhetik und Prothetik — Meilensteine in der Geschichte therapeutischer Hilfsmittel
Aesthetics and prosthetics — Milestones in the history of therapeutical appliances

Von Burkhard Jacob
By Burkhard Jacob

Seit einigen Jahren formiert sich im Design und in der Architektur ein neues Thema: Universal Design. Die Erwartungen sind hoch. Immer wieder wird gefordert, Probleme „nicht nur" behindertengerecht oder seniorenfreundlich zu lösen, sondern Produkte und Räume von Anfang an so zu gestalten, dass sie für alle Menschen leicht verständlich und mühelos zu benutzen sind, getreu der Devise: Was für Mütter mit Kindern, Menschen mit einem Handicap oder hochbetagte Senioren ein gestalterisches Muss ist, kann für andere Gruppen einen höheren Komfort bedeuten. Bei allen berechtigten Forderungen dürfen wir nicht ausblenden, was die Faszination von Ideen ausmacht, wie sie entstehen und sich entwickeln. Viele Dinge lassen sich erst erklären, wenn man sie als Evolution von Ideen versteht: Der Blick auf die Meilensteine der Ästhetik und Prothetik macht eins deutlich: Jede Form von etwas ist immer auch die Idee von etwas.

Dies zeigt insbesondere die Geschichte der Brille. Auch hier steht am Anfang ihrer Erfindung zunächst eine einfache Idee. Sie kommt allerdings nicht von den Ägyptern, Griechen oder Römern, sondern stammt von dem arabischen Gelehrten Abu Ali al-Hasan ibn al-Haitham, kurz Alhazen (965-1039 n. Chr.). In seinem Werk „Schatz der Optik" beschreibt er im 11. Jahrhundert die vergrößernde Wirkung einer plankonvexen Linse, einem Lesestein, der nicht auf das Auge, sondern auf die Schrift gesetzt wird und so die Buchstaben vergrößert.

Erst um 1240, als das Werk Alhazens nach Europa kommt und ins Lateinische übersetzt wird, verbreitet sich seine Idee in den Klöstern und bei den Gelehrten. Insbesondere die alterssichtigen und handwerklich begabten Mönche machen Gebrauch von der Entdeckung Alhazens und schleifen ihre Lesesteine aus Beryllkristallen, die der Brille ihren Namen geben. Es ist der Philosoph und Franziskanermönch Roger Bacon, der in seinem Werk „Opus Majus" von 1267 das Prinzip des Lesesteins erklärt und nach Verbesserungen sucht.

In recent years a new topic has been emerging in design and architecture: Universal Design. The expectations are high. Again and again the requirement is to find solutions to problems which not only fulfil the needs of disabled people or senior citizens, but instead to right from the start design products and spaces in such a way that they are easily understandable and effortless to use for all people, following the maxim: what is a must when it comes to design for mothers with small children, handicapped people and the elderly, can mean a higher degree of convenience for other groups. Despite all justified demands we must not forget what constitutes the fascination of ideas, how they evolve and develop. Looking at the milestones of aesthetics and prosthetics, makes one thing clear: every form of something is also always an idea of something.

This is demonstrated by the history of the spectacles in particular. The beginning of this invention is also marked by what is initially a simple idea. However, it does not come from the Egyptians, Greeks or Romans but from the Arab scholar Abu Ali al-Hasan ibn al-Haitham, or Alhazen for short (965-1039 AD). In his "Book of Optics" he describes the magnifying effect of a plano-convex lens, a reading stone, which is not placed in front of the eye but onto the writing instead, thus magnifying the letters.

It is not until 1240 when the work of Alhazen comes to Europe and is translated into Latin that his idea spreads through monasteries and among scholars. Presbyopic and manually skilled monks in particular make use of Alhazen's discovery and cut their reading stones out of beryl crystals (thus the German word for spectacles Brille). It is the philosopher and Franciscan monk Roger Bacon who in his "Opus Majus" of 1267 explains the principle of the reading stone and looks for enhancements.

The first spectacles are presumably manufactured in Italy in the 13th century. The sources are not clear on this. However, there are many

Die ersten Brillen werden vermutlich im 13. Jahrhundert in Italien hergestellt. Die Quellenlage ist hier nicht eindeutig. Es spricht aber vieles dafür, da Venedig als Wiege der mitteleuropäischen Glasherstellung angesehen werden kann. Bis 1295 werden alle Glasöfen Venedigs auf die Insel Murano verlagert, um das streng gehütete Geheimnis der Glasherstellung zu bewahren. Die Glasbläser sind zwar gut bezahlt, ihre Reisefreiheit wird aber per Gesetz eingeschränkt, damit sich ihr Wissen und ihre Ideen nicht verbreiten.

Die Sehhilfen des Mittelalters kommen noch ohne Bügel aus und ihre Linsen werden konvex geschliffen. Somit sind sie nur für die Korrektur der Weitsichtigkeit geeignet. Gleichwohl avanciert die Sehhilfe von Anfang an zum Statussymbol der gebildeten Schichten. Auch in künstlerischen Darstellungen tragen Philosophen und Propheten häufig eine Brille als Attribut ihrer Weisheit. Die Erfindung des Buchdrucks und die Verbreitung gedruckter Schriften verstärken dieses Bild und den Bedarf an Brillen. Insbesondere die Gelehrten und Mönche zeigen ein hohes Interesse an den mittelalterlichen Sehhilfen, erlauben sie es doch, Schriften bis ins hohe Alter zu lesen und zu verfassen.

Alters- und Fehlsichtige müssen allerdings noch bis ins 18. Jahrhundert die Sehhilfe in der Hand halten. Erst 1727 fertigt der englische Optiker Edward Scarlett erstmals ein Brillengestell. Und 1746 verkauft der französische Optiker Marc Mitouflet Thomin in Paris Brillengestelle mit seitlichen Bügeln, die am Kopf fixiert werden. Allerdings verursacht der Druck auf die Schläfen auf Dauer auch Kopfschmerzen. 1752 kommen aus England Brillen mit Gelenk und Ohrenbügeln, die den Tragekomfort verbessern. Im Jahr 1785 stellt kein Geringerer als Benjamin Franklin der Öffentlichkeit eine so genannte Bifokalbrille vor, bei der jeweils zwei Halblinsen in die Brillenfassung eingespannt sind. Die oberen Linsen dienen dem Sehen in die Ferne, die unteren Linsen dem Nahsehen. Und als der Pariser Optiker Poulot 1857 schließlich den Nasenhalter erfindet, entwickelt sich die Brille zu der Form, wie wir sie heute kennen. Seither hat sie sich kaum verändert und wird in ihrer Erscheinung im Wesentlichen durch neue Materialien und die Mode beeinflusst.

Im 20. Jahrhundert verbessert die wissenschaftliche Forschung die Brillenoptik. Insbesondere die Entdeckung des Punktalglases durch Louis Otto Moritz von Rohr im Jahr 1912 ermöglicht ein scharfes Sehen unabhängig von der Blickrichtung. Es folgen die Kunststoffgläser und die Entspiegelung, bis die deutsche Chemikerin Marga Faulstich im Jahr 1972 hochbrechende Brillengläser für eine hohe Dioptrienzahl entwickelt. Als langjährige Mitarbeiterin und Führungskraft der Firma Schott tragen viele Patente ihren Namen. Das von ihr entwickelte sehr leichte Glas reduziert nicht nur das Gewicht der Brillen um fast 40 Prozent, auch die dicken und unästhetischen Brillengläser gehören bald der Vergangenheit an.

indications that Venice can be regarded as the cradle of the middle European glass production. By 1295 all glass ovens in Venice are relocated to the island of Murano in order to protect the well-kept secret of glass production. The glass blowers are well paid, however, their freedom to travel is restricted by law to prevent their knowledge and ideas from spreading.

The visual aids of the Middle Ages do not have any arms and the lenses have a convex cut. Thus they are only suitable for correcting far-sightedness. Nevertheless, the visual aid becomes a status symbol of the educated classes right from the start. Also in artistic portrayals philosophers and prophets often wear spectacles as an attribute of their wisdom. The invention of the printing press and the spreading of printed texts reinforce this impression and increase the demand for spectacles. Scholars and monks in particular display great interest in the visual aids of the Middle Ages because they enable them to read and write texts into old age.

However, people with age-related long-sightedness and defective vision have to hold their spectacles in their hands until the 18th century. It is not until 1727 that the English optician Edward Scarlett manufactures the first ever spectacles frame. And in 1746 the French optician Marc Mitouflet Thomin sells spectacle frames in Paris which have side arms that are attached to the head. However, the pressure on the temples causes headaches in the long run. In 1752 spectacles with hinges and arms which increase wearing comfort are made in England. In 1785 no less a figure than Benjamin Franklin introduces the so-called bifocal spectacles to the public, which consist of two half lenses clamped into a spectacle frame. The upper lenses are for distant vision, the lower for near vision. And when the Parisian optician Poulot finally invents the nose holder in 1857, spectacles take on the form which we know today. Since then they have hardly changed and their appearance is essentially influenced by new materials and fashion.

In the 20th century scientific research enhances the spectacle optics. Particularly the invention of "Punktal" lenses by Louis Otto Moritz von Rohr in 1912 enables sharp vision independent of the direction of view. Plastic lenses and anti-reflective coatings are invented next before the German chemist Marga Faulstich develops highly refractive lenses for high dioptres in 1972. As a long-standing employee and executive of the company Schott many patents carry her name. The very light glass developed by her not only reduces the weight of spectacles by almost 40 per cent, but also thick and unaesthetic lenses will soon become a thing of the past.

Ästhetik Brille / Aesthetics Spectacles

Nachbildung eines Lesesteins. Replication of a reading stone.

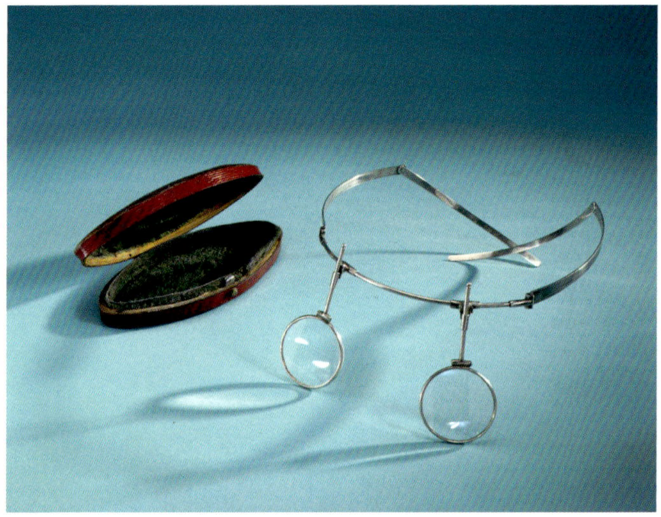

Stirnreifenbrille aus dem Jahr 1796. Headband spectacles, 1796.

1100 1200 1300 1400 1500 1600

Im 11. Jahrhundert n. Chr. beschreibt der arabische Gelehrte Alhazen in seinem Werk „Schatz der Optik“ erstmals die vergrößernde Wirkung einer plankonvexen Linse.
11th century AD: The Arabic scholar Alhazen describes the magnifying effect of a plano-convex lens in his “Book of Optics” for the first time.

1267

erklärt der Philosoph und Franziskanermönch Roger Bacon in seinem Werk „Opus Majus“ das Prinzip des Lesesteins und sucht nach Verbesserungen.
The philosopher and Franciscan monk Roger Bacon explains the principle of the reading stone and looks for enhancements in his book “Opus Majus”.

1352

porträtiert Tommaso di Modena Kardinal Hugo de Province mit einer Nietbrille.
Tommaso di Modena paints a portrait of Cardinal Hugo de Province with rivet spectacles.

Ohren- oder Schläfenbrille, 1800-1802. Transverse folding spectacles, 1800-1802.

Nasenkneifer (Brille ohne Bügel) mit Etui, 1870-1920.
Pince-nez spectacles with case, 1870-1920.

1700 1800 1900

1727

verbindet der englische Optiker Edward Scarlett das Brillengestell mit Ohrenbügeln.

The English optician Edward Scarlett combines the spectacle frame with ear temples.

1746

verkauft der französische Optiker Marc Mitouflet Thomin in Paris Brillengestelle mit seitlichen Bügeln.

The French optician Marc Mitouflet Thomin sells spectacles with side arms in Paris.

1785

fertigt der amerikanische Staatsmann und Wissenschaftler Benjamin Franklin die erste Bifokalbrille für das Sehen in der Ferne und in der Nähe.

The American statesman and scientist Benjamin Franklin produces the first bifocal glasses for distant and near vision.

1857

erfindet der Pariser Optiker Poulot den Nasenhalter.

The Parisian optician Poulot invents the nose holder.

1912

entwickelt der deutsche Optiker Louis Otto Moritz von Rohr die Punktalbrillengläser, die unabhängig von der Blickrichtung scharfe Bilder geben.

The German optician Louis Otto Moritz von Rohr develops the "Punktal" lenses which provide sharp images independent of the direction of view.

1972

entwickelt die deutsche Chemikerin Marga Faulstich hochbrechende, sehr leichte Brillengläser für eine hohe Dioptrienzahl.

The German chemist Marga Faulstich develops very light, highly refractive lenses for high dioptres.

Ästhetik Prothetik / Aesthetics Prosthetics

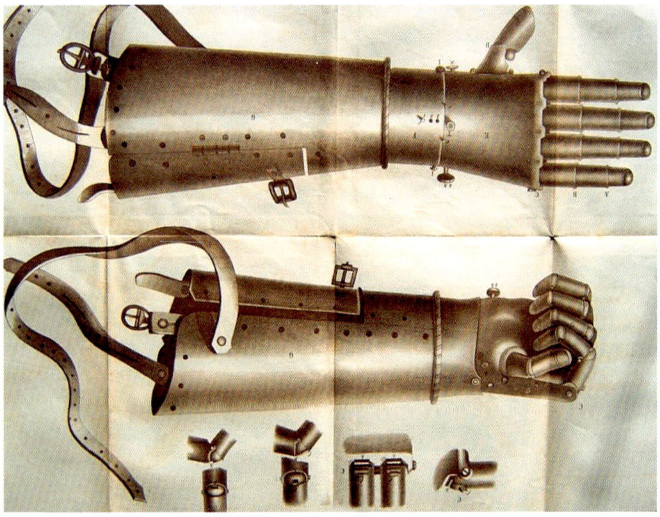

„Eiserne Hand“ des deutschen Ritters Götz von Berlichingen, 1504-1510.
German imperial knight Götz von Berlichingen's “Iron Hand”, 1504-1510.

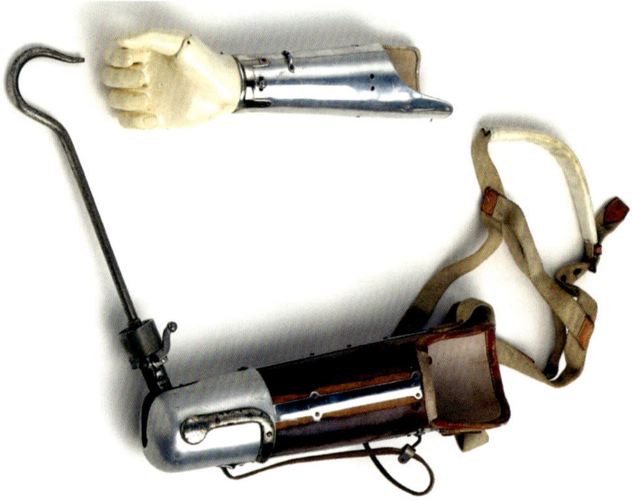

Künstlicher Arm mit Haken und Handprothese, 1924-1930.
Artificial arm with hook and forearm attachments, 1924-1930.

1500 1600 1700 1800

1504
verliert der deutsche Ritter Götz von Berlichingen in einer Schlacht bei Landshut seine rechte Hand. Er ersetzt sie durch eine eiserne Prothese aus mehr als 200 Einzelteilen.
In a battle near Landshut, the German knight Götz von Berlichingen loses his right hand. He replaces it with an iron prosthesis consisting of more than 200 individual parts.

1561
veröffentlicht der französische Militärarzt Ambroise Paré erstmals eine ausführliche Schrift über Prothesen und die menschliche Anatomie.
The French military doctor Ambroise Paré publishes the first detailed text about prosthetics and the human anatomy.

1812
entwickelt Pierre Ballif eine Hand aus Messingblech, die nur noch 500 g wiegt.
Pierre Ballif develops a hand out of sheet brass which weighs only 500 grammes.

1832
veröffentlicht die Berliner Instrumentenmacherin Margarethe Karoline Eichler Normen für den Bau von Kunstbeinen.
The Berlin toolmaker Margarethe Karoline Eichler publishes standards for the construction of artificial legs.

1836
konstruierte Eichler einen verbesserten Kunstarm aus Neusilberblech, der nur noch 250 g wiegt.
Eichler creates an improved artificial arm made of sheet nickel silver which weighs only 250 grammes.

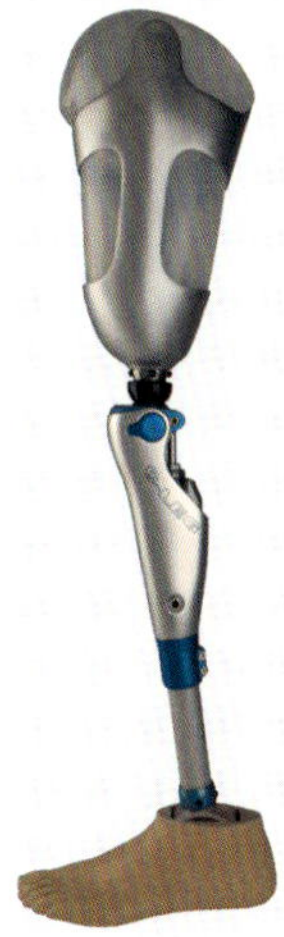

C-Leg von Otto Bock: das erste vollständig mikroprozessorgesteuerte Kniegelenk. Otto Bock's C-Leg is the first fully microprocessor-controlled knee joint.

DynamicArm von Otto Bock: erstes elektronisch gesteuertes Elektro-Ellbogengelenk. Otto Bock's DynamicArm is the first electronically controlled elbow joint.

1900

1860

legt der Pariser Graf de Beaufort Ersatzglieder für Ober- und Unterarme vor, die nur 25 Francs kosten, zu Recht eine „Prothese du pauvre".
The Parisian Count de Beaufort presents replacement limbs for upper and lower arms, which cost only 25 francs; they are appropriately called "prothese du pauvre".

1860

entwickelt der französische Instrumentenmacher Joseph-Frédéric-Benoît Charrière eine Prothese mit ledernen Hülsen, die samt Hand- und Fingergliedern aus Stahl und Holz nur 350 g wiegen.
The French toolmaker Joseph-Frédéric-Benoît Charrière develops a prosthesis with leather arm shells which weighs only 350 grammes including the hand and finger phalanges made of steel and wood.

1898

publiziert der Italiener Giuliano Vanghetti seine Idee einer „cinematischen Prothese". Er nutzt die Muskeln und Sehnen im Amputationsstumpf für die Bewegung der künstlichen Hand.
The Italian Giuliano Vanghetti publishes his idea of a "cinematic prosthesis". He uses the muscles and tendons in the amputation stump to operate the artificial hand.

1911

führt der amerikanische Mechaniker William T. Carnes die Elemente früherer Konstruktionen zu einem neuen Kunstarm und zu einer neuen Kunsthand zusammen.
The American mechanic William T. Carnes combines elements of earlier constructions to create a new artificial arm and a new artificial hand.

1915

schuf der deutsche Chirurg Ernst Ferdinand Sauerbruch die chirurgischen und mechanischen Voraussetzungen für eine willkürlich bewegliche künstliche Hand, ohne auf die Hilfe der gesunden Hand angewiesen zu sein.
The German surgeon Ernst Ferdinand Sauerbruch creates the surgical and mechanical prerequisites for a voluntarily movable artificial hand, which does not require help from the healthy hand.

Medizin kaufen — Über einen blinden Fleck der Kostendebatte im Gesundheitswesen
Buying medicine — About a blind spot in the healthcare cost debate

Von Prof. Dr. Birger P. Priddat
By Prof. Dr. Birger P. Priddat

Kranke konsumieren Medizin, um gesund zu werden. Sie gehen zur Apotheke, um sich Medizin oder Hilfsmittel zu kaufen, von denen sie gelesen oder gehört haben, sie machen gesund. Sie gehen zu Ärzten, die ihnen die Medizin oder Hilfsmittel verschreiben, die sie bei der Apotheke abholen. Der Kauf und Konsum von Medizin bzw. der Gebrauch von Hilfsmitteln gibt noch keine Gewähr, dass man gesund wird. Beide Medizinkonsummärkte sind separiert; im ersten Fall bezahlen die Menschen selber, im zweiten Fall zahlt ihre Kranken- bzw. Pflegeversicherung – der sie zuvor Beiträge bezahlt haben.

Da die Kosten der Krankenbehandlung steigen, fragen die Krankenkassen, welche Medizin oder Hilfsmittel wirksamer sind als andere und zu welchem Preis. Um dies nachzuweisen, ist eine neue Wissenschaft entstanden: die „Pharmaco-economics", die Pharmako-Ökonomie.

Die Pharmako-Ökonomie ist zunächst einmal nichts anderes als eine Übertragung gewöhnlicher ökonomischer Kosten-Nutzen-Betrachtungen auf besondere Produkte: auf Pharmazeutika und therapeutische Hilfsmittel. Dabei wird eine angenommene Wirksamkeit beziehungsweise Effektivität des Produktes gegen die Herstellungskosten gerechnet – das sind die direkten Kosten der Produktion inklusive der Forschung und des Marketings –, gegen die direkten Kosten der Therapie und gegebenenfalls gegen indirekte volkswirtschaftliche Kosten wie etwa den Produktivitätsverlust des Kranken. Dies gilt gleichermaßen für Arzneimittel wie für therapeutische Hilfsmittel, zum Beispiel Rollstühle, Mobilitätshilfen, Orthopädietechnik, Prothesen, Physiotherapie, Ergotherapie, ambulante Pflege, Pflegedienste, -geräte und -mittel.

Dieses Verfahren wird als effizientes Instrument zur Behebung der aktuellen Kostenkrise im Gesundheitswesen angesehen. Zielkriterium ist im einfachsten Fall ein Optimierungskalkül, wonach aus einer

Persons suffering from a disease consume pharmaceuticals in order to get well. They go to a pharmacy to buy medicine or therapeutic aids they have heard of or read about that they would help them recover. They go to physicians prescribing them pharmaceuticals or therapeutic aids which they then pick up at the pharmacy. However, purchase and consumption of medicine and/or use of therapeutic aids are no guarantee for becoming well again. The two medical consumption markets are separated; in the first case, people pay the items from their own money, in the second case, their health and/or nursing care insurance pays the bill – after the insurant has paid the insurance premium fee.

As the costs for healthcare are rising, healthcare insurance companies are inquiring into what specific pharmaceutical or therapeutic aid would be more effective than others, and at what price. To provide evidence of this, a new science has come into being: the so-called "pharmacoeconomics".

Pharmacoeconomics is, at first, nothing else than the transfer of ordinary cost-benefit considerations to special products: to pharmaceuticals and therapeutic appliances. In doing so, an assumed efficacy and/or efficiency of the product is calculated against the costs of production – these are the direct costs of manufacture including expenditures for research and marketing –, against the direct costs of therapy and, as the case may be, against indirect economic costs such as the loss in productivity of the diseased person. This applies to both pharmaceutical products and therapeutic appliances, for example wheelchairs, mobility aids, orthopaedic techniques, prostheses, physiotherapy, ergotherapy, ambulatory care, as well as nursing services, devices, and products.

This method is considered an efficient instrument for dealing with the current cost crisis in the healthcare system. In the simplest case, the target criterion is an optimisation consideration according to which

GESUNDHEITSAUSGABEN HEALTH EXPENDITURE

Gesundheitsausgaben pro Kopf in US-Dollar, OECD-Länder
Health expenditure per capita in US-Dollar, OECD countries

2007

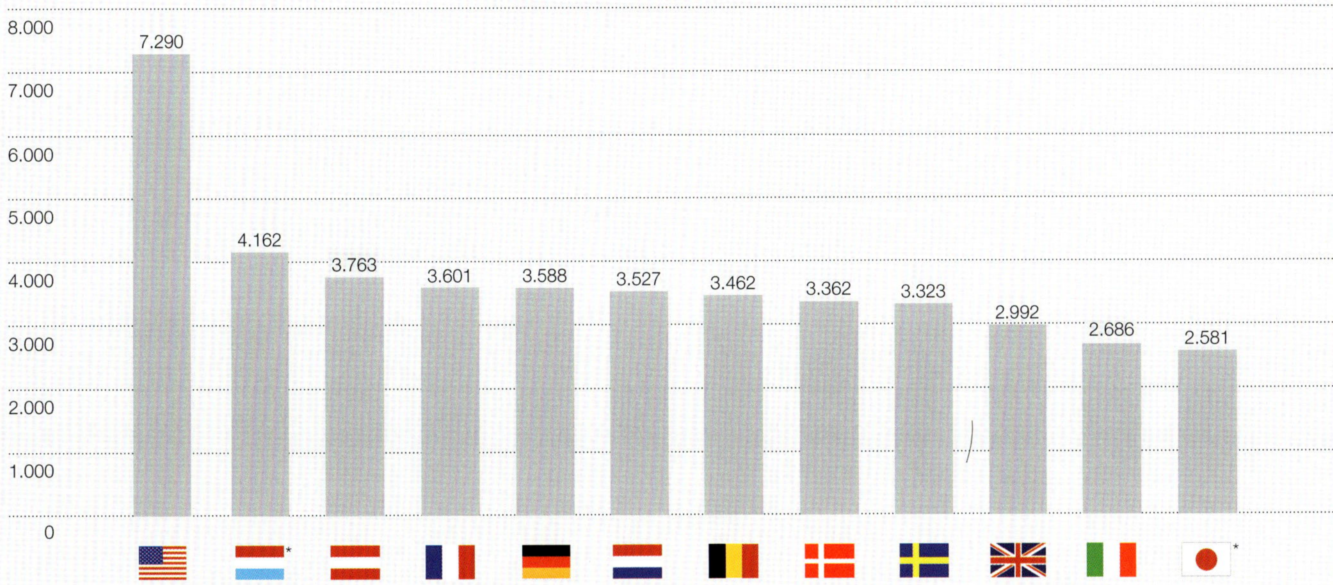

* 2006
Quellen: OECD, Organisation für wirtschaftliche Zusammenarbeit und Entwicklung, 2009
Sources: OECD, Organisation for Economic Co-operation and Development, 2009

Menge von Therapeutika gleicher Wirksamkeit dasjenige mit den geringsten Herstellungskosten verschrieben und konsumiert werden soll (Kostenminimierungsanalyse); andere Analyseformen sind komplexer, aber auch mit größeren Bewertungs- und Informationsproblemen behaftet.

Die „Kosteneffektivität" von Arznei- und Hilfsmitteln zum Beispiel, die in vielen „pharmako-ökonomischen" Studien analysiert wird, sagt zu wenig über die tatsächliche Wirksamkeit der Arzneien und Hilfsmittel aus. Sie kann, wie etwa Brennecke und Versteegen eingewandt haben, durch das Verschreibungsverhalten und die wissenschaftlichen Kenntnisse des Arztes, aber auch durch den individuell verschiedenen Krankheitszustand oder die Therapietreue des Patienten stark beeinflusst werden. Erst in der ärztlichen Verschreibungspraxis und ihrer Fall-zu-Fall-Auswertung wird also sichtbar, welche tatsächlichen Wirksamkeiten neue Hilfsmittel und Pharmazeutika haben. Jedes Therapeutikum – aus einer durchschnittlichen Menge von zum Beispiel zwanzig bis dreißig ähnlichen auf dem Markt angebotenen Mitteln – hat spezifische Wirkungen je nach Alter, Geschlecht, Krankheitsverlauf, Gesamtdisposition, Therapieanweisungen und so fort. Die pharmazeutisch behauptete Wirkung wiederum ist eine ideale Durchschnittswirkung, die über die effektive Wirksamkeit eines neuen Mittels nichts aussagt.

Es fehlen also Bewertungen der Qualität von Behandlungsergebnissen, aufgrund derer erst Kostenvergleiche möglich wären. Dieser – durch die üblichen Studien nicht behobene – Mangel an Wissen um die effektive Wirksamkeit neuer Arznei- und Hilfsmittel ist selber kostenintensiv: Fehldiagnosen, Fehlbehandlungen, die Vernachlässigung von Nebenwirkungen und dergleichen mehr summieren sich zu einem beachtlichen Faktor bisher noch kaum beachteter „Einführungs-Versuchs-Kosten". Grund dafür sind vornehmlich die Wirkungslosigkeit (Krankheitsverlängerung für die Patienten, Produktion neuer Diagnose- und Therapiekosten und so fort) beziehungsweise die Erzeugung neuer Erkrankungen (Nebenwirkungen, Verschlechterungen der Ausgangserkrankungen) – die Belastung für die Patienten noch gar nicht gerechnet.

Die Kosten der Fehldiagnose

Wie sind die Kosten von Fehldiagnose und Fehltherapie zu bewerten? Um etwaigen Missverständnissen gleich zu begegnen: Wer so fragt, dem geht es nicht um Schuldzuweisungen an die Ärzte, sondern um eine praxisgerechte Analyse. Die Produktion und das Marktangebot von Arznei- und Hilfsmitteln sind tatsächlich nur ein Therapieangebot, noch keine Therapie. Erst in der Hand der Ärzte bekommt das Medikament Wirkung; aber wie, wenn die Ärzte über die Wirkungsspezifität ähnlicher Therapeutika nichts wissen (der Fall der „unvollkommenen Information") oder wenn sie unter Stressbedingungen ärztlicher Praxis Fehldiagnosen stellen und damit Fehltherapien verfügen?

the therapeutic agent with the lowest production costs is selected out of a range of other therapeutics with the same effectiveness, and which is supposed to be prescribed and consumed (cost minimisation analysis); other forms of analysis are more complex but also afflicted with major assessment and information problems.

The "cost effectiveness" of pharmaceuticals and therapeutic aids, for example, which is analysed in many "pharmacoeconomic" studies, does not provide sufficient information about the actual effectiveness of these pharmaceuticals and therapeutic aids. As Brennecke and Versteegen have pointed out, it can be strongly influenced by the physician's prescription behaviour and his scientific knowledge, but also by the patients' individually different conditions of illness or therapy adherence. As such, the actual effectiveness of new aids and pharmaceuticals only becomes visible in the doctors' prescription practice and in case-by-case assessments. Each therapeutic agent – from a range of, for instance, twenty to thirty similar agents offered in the market – has specific effects, depending upon age, gender, course of disease, overall disposition, therapy instructions, and so on. The pharmaceutically alleged effect, in turn, is an "ideal average effect" which does not state anything valid about the actual effectiveness of a new therapeutic agent.

So what is missing are quality assessments of therapy results which would then, and only then, allow correct cost comparisons. This lack of knowledge about the actual effectiveness of new therapeutic agents and aids – not resolved by conventional studies – is in itself a major cost-intensive factor: misdiagnosis, mistreatment, neglect of undesirable side effects and the like sum up to a considerable factor of what could be called "introductory experimental costs" – hardly taken into account so far. The primary reasons for this are ineffectiveness of agents (prolongation of the patient's illness, production of new diagnosis and therapy costs, etc.) and/or production of new indispositions (side effects, worsening of the original disease situation) – not even taking the stresses and strains for the patients into account at this point.

The costs of misdiagnosis

How can the costs of misdiagnosis and mistreatment be estimated? To avoid any misunderstandings right here and now: it is not the intention of those who ask this question to put the blame on physicians; the only thing they are concerned with is a practice-oriented analysis. Production and market offer of pharmaceuticals and therapeutic aids are actually only a therapy offer, not already a therapy in itself. The therapeutic agent only becomes effective in the hands of physicians; the question, however, is how, when the medical practitioner does not know anything about the specific effects of similar therapeutic agents (the case of "insufficient information"), or when they, under the stress and pressure of daily medical practice, misdiagnose specific symptoms and thus embark upon erroneous therapies?

EUROPA / EUROPE

MARKT FÜR AUSGEWÄHLTE MEDIZINISCHE HILFSMITTEL*

MARKET FOR SELECTED MEDICAL APPLIANCES*

Einfuhr und Ausfuhr in Millionen Euro

Import and export in million euros

2008

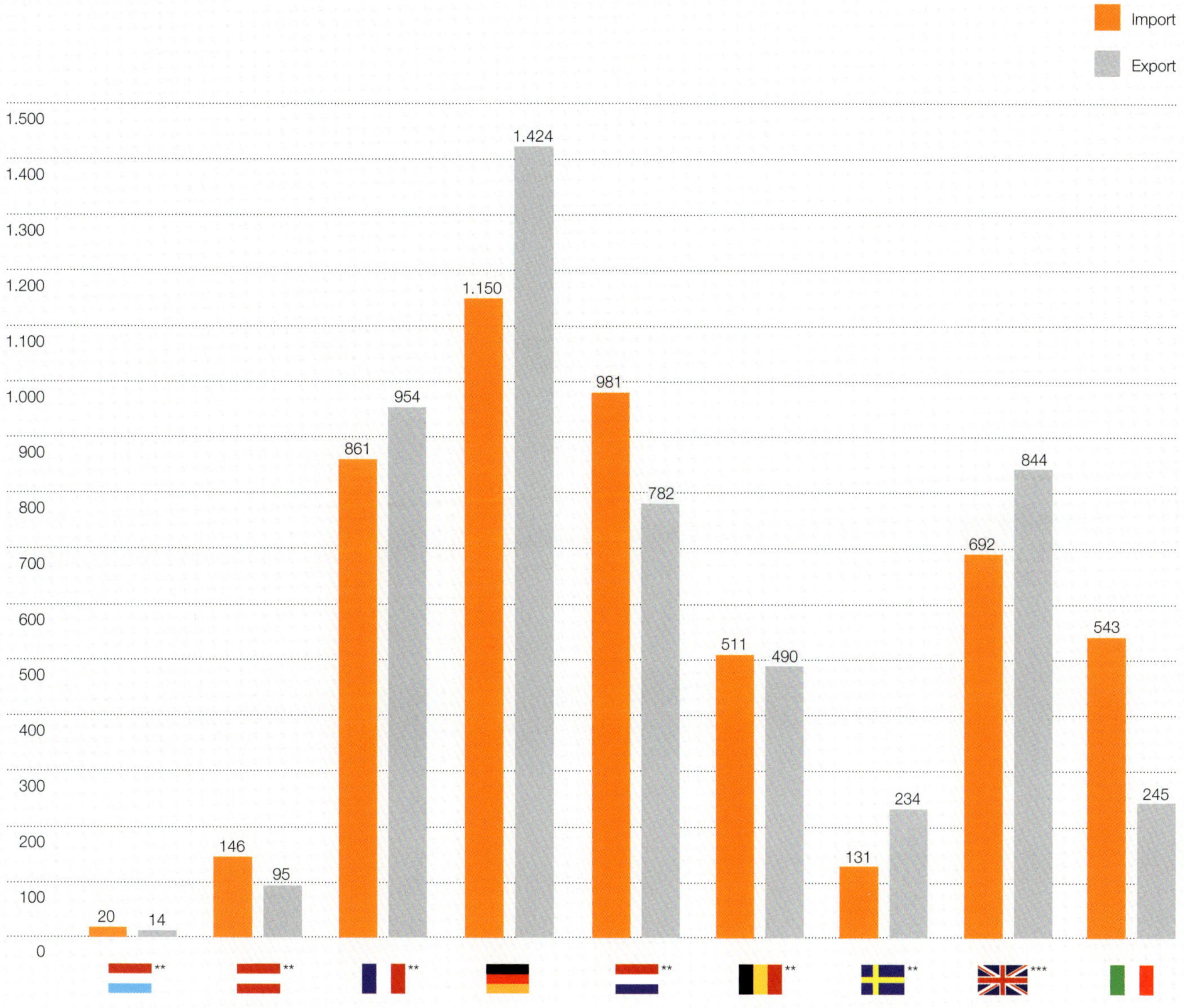

* Ausgewählte Produkte: Rollstühle mit/ohne Motor, Apparate für Atmungstherapie, orthopädische Vorrichtungen, Schienen für Knochenbrüche, künstliche Gelenke, Prothesen und Implantate

* Selected products: Wheelchairs with/without motor, devices for breathing therapy, orthopaedic appliances, splints for bone fractures, artificial joints, prostheses and implants

** Januar – November 2008 / January – November 2008

*** Januar – Oktober 2008 / January – October 2008

Quellen / Sources:
Germany Trade & Invest, Berlin,
Statistisches Bundesamt, Wiesbaden / Federal Statistical Office, Wiesbaden

Erst eine genauere Untersuchung der – um einen ökonomischen Ausdruck zu benutzen – Gutsspezifität der einzelnen Pharmazeutika in Bezug auf eine Symptomgruppe kann Informationen beschaffen, die Ärzten eine höhere Diagnosegenauigkeit erlauben. Erst dann ließen sich auch die hohen Transaktionskosten des Experimentierens mit verschiedenen Fehlbehandlungen und Nebenwirkungen senken, die diesen Transaktionskosten zugeschlagen werden müssen.

Wenn man ökonomisch an die Therapeutika herangeht, ist nicht der Verschreibungsakt entscheidend, sondern der effektive Konsum des Arzneimittels, das heißt die wirksame Behandlung. Aus Sicht der Hersteller genügt der Verschreibungs- beziehungsweise Verkaufserfolg. Aus Sicht der Medizin und ihrer Institutionen kann aber nur der Therapieerfolg zählen. Freilich rechnen diese klinischen Gesichtspunkte nicht den Aufwand ein, der nötig ist, bis tatsächlich eine krankheitsregulierende Wirkung der Medikamente eintritt. Dieser Aufwand lässt sich in den Transaktionskosten des Therapieerfolges bemessen. Brian Dickson von der Princeton University listet folgende – transaktionskostenrelevante – Faktoren der Therapeutika auf: Verursacht das Medikament Nebenwirkungen? Wie effektiv ist es? Wie schnell wirkt es? Wie leicht ist es zu handhaben?

Wenn ich hier den Begriff der Transaktionskosten benutze, dann meine ich damit dies: Ökonomisch gesehen ist der Arzt nur eine Art Zwischenhändler zwischen Pharmazie- und Therapieherstellern und dem Patienten, seinem Kunden. Da er Medikamente und Therapien unter Berücksichtigung medizinischer Qualitätskriterien verschreibt, definiert er durch seine medizinische Beratertätigkeit die tatsächliche Nachfrage. Für die Patienten ist er eine Vertrauensperson, der sie sich therapeutisch vollständig überlassen. Arzt und Patient handeln also gleichsam einen impliziten Vertrag aus, in dem Beratungs- beziehungsweise diagnostisch-therapeutische Kompetenz gegen Behandlungserfolg getauscht wird. Dieser Vertrag ist deshalb „implizit", weil er sich vom expliziten Vertrag zwischen Patienten und Arzt beziehungsweise den Kassen unterscheidet: Dort wird nur die Behandlung bezahlt, nicht aber der Erfolg.

Der implizite Vertrag wird unter der Unsicherheit abgeschlossen, dass beide Vertragsparteien nicht wissen können, ob sie das „liefern" können, was vereinbart wurde. „Liefern" sie beide, was vereinbart wurde – der Arzt seine Kompetenz, der Patient die Heilung –, fallen keine Transaktionskosten an. Alle anderen Zustände – Fehlbehandlungen, Wirkungslosigkeit der Pharmazeutika und therapeutischen Hilfsmittel, Nebenwirkungen – bilden in ihrer Summe die Transaktionskosten dieses impliziten Vertrages. Sie steigen, je weniger die Ärzte wissen, wie oder bei wem die Therapeutika, die sie verschreiben, tatsächlich anschlagen können.

Transaktionskosten entstehen, wenn zwischen der Wirkungszuschreibung des Medikaments und der tatsächlichen Wirkung Differenzen bestehen, die keine therapeutischen Standardlösungen mehr zulassen oder die es erfordern, dass die therapeutischen Standards hoch ausdifferenzieren werden. Ökonomisch spricht man dann von

Only a detailed examination of the – to use a term from the field of economics – "effect specifity" of the individual pharmaceuticals with respect to the group of symptoms can provide information allowing practitioners a higher degree of diagnosis accuracy. Then, and only then, can the high transaction costs of experimenting with various mistreatments and side effects, which have to be added to these transaction costs, be reduced.

When approaching therapeutic agents from an economical point of view, the crucial factor is not the act of prescription but the effective consumption of the therapeutic agent, that is, effective treatment. From the manufacturers' point of view, the success of prescription and/or sale of the therapeutic agent suffices. From the physician's point of view, and that of their institutions, the only thing that counts is successful treatment. True, these clinical viewpoints do not factor in the effort that is necessary until a disease-regulating effect brought about by the therapeutic agents actually takes place. This effort can be measured by the transaction costs of therapeutic success. Brian Dickson of Princeton University lists the following – transaction cost relevant – factors of therapeutic agents: does the agent cause any side effects? How effective is it? How fast does it act? How easy is it to administer?

When I use the term "transaction costs" here, this is what I mean: from an economic point of view, the physician is only a kind of "intermediary" between producers of pharmaceuticals and therapy providers on the one side, and the patient, his client, on the other. As he prescribes pharmaceuticals and therapies with consideration to medical quality criteria, he defines the actual demand by means of his activities as medical consultant. In his patients' view, he is a confidant they fully trust when it comes to their therapy. In this sense, physician and patient engage in an "implicit" contract in which the physician's competence in terms of consultation, diagnosis, and therapy is exchanged against successful treatment. This contract is "implicit" because it differs from the explicit contract between patient/practitioner and the health insurance companies: the only thing that is paid for there is the treatment but not the success.

This implicit contract is "closed" bearing the uncertainty in mind that both contractual parties do not know whether they can "deliver" what was agreed upon. If they both "deliver" what was agreed upon – the physician by means of his competence, the patient by means of his recovery – no transaction costs are incurred. All other conditions – mistreatment, ineffectiveness of pharmaceuticals and therapeutic aids, side effects – amount to the sum total of transaction costs of this implicit contract. They rise to the degree that practitioners do not know how, and on whom, the therapeutic agents prescribed by them have an actual effect.

Transaction costs are incurred when there are differences between the effects ascribed to the pharmaceutical agent and its actual effects that do not allow for standard therapeutic solutions, or which

DEUTSCHLAND / GERMANY

IMPORTE UND EXPORTE AUSGEWÄHLTER MEDIZINISCHER HILFSMITTEL*

EXPORT AND IMPORT OF SELECTED MEDICAL APPLIANCES*

Importe und Exporte in Millionen Euro

Import and export in million euros

2003 – 2008

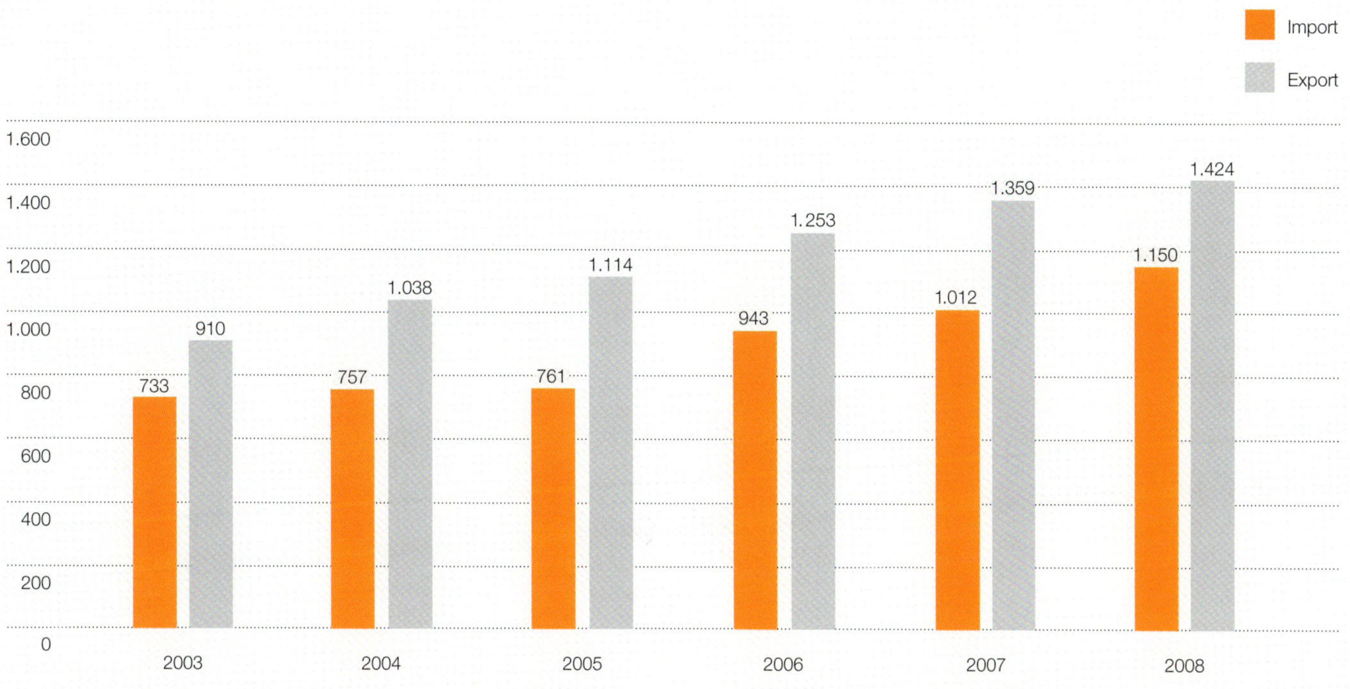

* Ausgewählte Produkte: Rollstühle mit/ohne Motor, Apparate für Atmungstherapie, orthopädische Vorrichtungen, Schienen für Knochenbrüche, künstliche Gelenke, Prothesen und Implantate

* Selected products: Wheelchairs with/without motor, devices for breathing therapy, orthopaedic appliances, splints for bone fractures, artificial joints, prostheses and implants

Quelle / Source:
Statistisches Bundesamt, Wiesbaden / Federal Statistical Office, Wiesbaden

den Kosten der Realisierung des Vertrages, eben von Transaktionskosten. Der Patient ist in diesem Vertrag mit dem Arzt im Vertrauen darauf eingegangen, dass dieser weiß, was er tut. Da der Arzt aber gemeinhin gar nicht genau wissen kann, was er tut, wenn er ein Medikament verschreibt, drohen Therapieversagenskosten, die bisher noch nicht ernsthaft berücksichtigt wurden. Um sie berücksichtigen zu können, bräuchte es allerdings neue Analyseverfahren: Es bedarf zum Beispiel der „Therapie-Wirkungsforschung (qualitative Beschreibung von Gesundheitseffekten) sowie einer Epidemiologie des Nutzens von Arzneimitteln (quantitative Erfassung von Verteilung und Auftretenshäufigkeit von Gesundheits-Effekten)".

Transaktionskosten sind keine imaginären Kosten. Man kann sie in Geld ausdrücken: Die Unwirksamkeit bestimmter Therapeutika „kostet" Doppel- und Mehrfachmedikamentation, zugleich hält die Krankheit länger an, höhere Kosten durch Ausfall der Arbeit sowie längere und teurere Behandlungszeiten entstehen, unerwartete Nebenwirkungen bei Fehlbehandlungen erfordern zusätzlichen medizinischen Aufwand und so weiter. Auch sinkt das „Vertrauenskapital", das die Patienten den Ärzten gegenüber haben. Folge ist oft der Wechsel des Arztes, also die Kündigung des impliziten Vertrages. Jeder Arzt wird etliche einschlägige Fälle kennen. Dennoch gibt es über diese Kosten natürlich noch keine Statistik, weshalb man vorerst nur Fall-zu-Fall-Berechnungen machen kann. In den Transaktionskosten unsicherer Arzt-Patient-Beziehungen wird mithin ein höchst ineffizienter Bereich des Gesundheitswesens sichtbar, der unter dem Mantel der hehren ärztlichen Kunst verborgen bleibt und die diagnostische und therapeutische Überlastung von Ärzten zutage treten lässt. Die Überforderung vieler Ärzte – welches Therapeutikum für welche Diagnose? – berührt allerdings auch eine andere ökonomische Beziehung: die zwischen pharmazeutischen und Hilfsmittelherstellern und Ärzten. Welche Produktinformationen erhalten die ärztlichen „Zwischenhändler" eigentlich?

Die schlichte „Pharmako-Ökonomie", die allein das kostengünstigere Medikament oder Hilfsmittel im Auge hat, führt offenbar in die therapeutische Irre. Der Preis kann nicht die Information über die therapeutische Effektivität ersetzen. Nicht das „billigere" Mittel ist das ökonomisch „beste", sondern das medizinisch effektivste. Hier muss die Ökonomie in Qualitäten denken lernen. Sie muss informatorische Leistungen fordern, die eine Kostensenkung durch präzises Wissen um die je spezifischen Wirksamkeiten erzeugen. Man kann es auch anders formulieren: Es bedarf patientengerechter Wirkungsbeschreibungen, die bisher kaum erforscht sind. Erst solche Produktdifferenzierungen würden jedoch eine lebensqualitätsgerechte Therapie zulassen.

require a high degree of differentiation of therapeutic standards. From an economic point of view, these are the costs of realising the contract, namely the transaction costs. The patient has "closed" the contract with the physician trusting that the latter knows what he is doing. Yet, as the physician cannot know precisely in each case what he is doing when he prescribes a specific pharmaceutical, the threat of costs for therapy failures arises which so far couldn't be taken into account in earnest. In order to take them into account, however, new analysis methods would be required: for example, there is a need for "therapy effectiveness research (qualitative description of health effects) as well as an epidemiology of the benefit of pharmaceuticals (quantitative compilation of the distribution and frequency in the appearance of health effects)".

Transaction costs are no imaginary costs. They can be expressed in terms of money: the ineffectiveness of certain therapeutic agents results in costs for double and multiple medication; at the same time, the illness lasts longer, higher costs are also incurred by loss of work productivity of the patient; longer and more expensive treatment periods are necessary; unexpected side effects as a result of mistreatment require additional medical efforts, and so on. Moreover, the "trust capital" invested by the patients in their physicians is declining. Frequently, the consequence is the patient's change of physicians, that is, the "cancellation" of the implicit contract. Nearly every medical practitioner will be familiar with quite a number of pertinent cases. Even so, there are of course no relevant statistics to that degree yet, so that, for the time being, one can only resort to case-by-case calculations. Hence, a highly inefficient section of the healthcare system becomes visible in the transaction costs of uncertain practitioner/patient relationships which stay concealed under the cloak of noble Aesculapian art, and which brings to light the stresses and strains experienced by practitioners in terms of diagnosis and therapy. However, these stresses and strains many of them face – which therapeutic agent for which diagnosis? – touch on another economical relationship: the one between producers of pharmaceuticals and therapy providers on the one hand, and medical practitioners on the other. What product information do the medical "intermediaries" actually receive?

Mere "pharmacoeconomics" – only placing emphasis upon the more cost-effective therapeutic agent or aid – obviously puts the therapy on the wrong track. The price of the drug cannot replace the information about its specific therapeutic effectiveness. Not the "cheaper" pharmaceutical is the "best" in an economic sense, but the most effective one in a medical sense. This is the point where economics has to learn to think in terms of quality. It has to demand informational services bringing about cost reduction by precisely knowing about specific effects in each case. Or, to word it differently: there is a need for patient-oriented descriptions of medical effects that have hardly been researched so far. Only product differentiations of that nature would allow for therapies that will do justice to the patient's quality of life.

Zusammentreffen von Ethik und Ökonomie

Die Senkung der Belastung der Patienten durch unwirksame oder Fehlbehandlungen geht einher mit der Kostenreduktion durch spezifizierbare Pharmazeutika- und Hilfsmittelverwendung – ein überraschendes Zusammentreffen von Ethik und Ökonomie. Ethisch ist zunächst der tatsächliche Nutzen von Pharmazeutika und Hilfsmitteln für Patienten, nicht der funktionale Nutzen der Hersteller, ein Produkt auf den Ärzte- und Verschreibungsmarkt gebracht zu haben. Natürlich ist es ökonomisch effizient, ein Produkt „vermarktet" zu haben. Aber die erhöhte Verschreibungsrate von Ärzten für ein Produkt ist nur ein Teilmarktergebnis. Erst der tatsächliche Konsum entscheidet über die Produktqualität. Erst wenn wir vom Patienten und seiner Bedürfnisbefriedigung her denken, haben wir die ökonomische Analysekette vollständig geschlossen, weshalb die Erörterung ethischer Fragen – wie üblich – separat von den Wirtschaftlichkeitsuntersuchungen erfolgen kann.

Wenn wir durch wirkungsspezifischere pharmazeutische Informationen kürzere Behandlungsdauern (identisch mit kürzeren Krankheitsdauern) erreichen, wenn die „trial and error"-Methodik der Ärzte einer genaueren Therapieentscheidung weichen kann, wenn damit die Fehlbehandlungsrate sinkt (und damit längerfristig das Image der allopathischen Medizin wieder steigen kann), dann wird die blinde „Kosten-Effektivität" der heute noch dominanten „Pharmako-Ökonomie" einem differenzierten Kosten-Nutzen-Bild weichen können. In diesem Bild wird nicht mehr die Senkung der Herstellungskosten eines Medikaments allein ausschlaggebend sein, sondern die der gesamten Kosten, die für den Therapieerfolg nötig sind – einschließlich der Transaktionskosten der Arzt-Patient-Therapiebeziehung. Die Transaktionskosten der Arzt-Patient-Vertragsbeziehung zu senken ist ein bislang fast unbekannter Reduktionsfaktor – ein blinder Fleck in der pharmako-ökonomischen Diskussion. Der Pharmazeutikamarkt und der Markt für therapeutische Hilfsmittel ist, selbst für Ärzte, intransparent geworden. Wir benötigen daher sehr viel genauere, hoch ausdifferenziertere Produktinformationen. Das Produktmarketing der Hersteller wiederum kann nicht mehr allein unter bloßen betriebswirtschaftlichen Marketinggesichtspunkten betrachtet werden, sondern muss sich bewusst werden, dass hier die entscheidenden therapeutischen Informationen zu liefern sind, die die gesamtwirtschaftlichen Kosten senken, indem sie die Ärzte den Patienten treffgenauer helfen lassen.

Encounter between ethics and economy

Reducing the burden for the patient suffering from ineffective or mistreatment goes hand in hand with cost reduction by means of a specifiable use of pharmaceuticals and therapeutic aids – a surprising concurrence of ethics and economy. Ethical is, first of all, the actual benefit of pharmaceuticals and therapeutic aids for patients, not the functional benefit for the manufacturer of having brought a product to the physician and prescription market. Of course it is economically efficient to have "marketed" a product. But the higher prescription rate for a product is only a submarket result. Only the actual consumption determines the quality of a product. It is only when we start to think from the patient's point of view and the satisfaction of his needs that we have fully completed the economical analysis chain; for this reason, the consideration of ethical questions can take place – as usual – separate from economic feasibility studies.

If we manage to reduce the time spent for treatment and therapy (and thus the duration of a disease), if the physicians' "trial and error" approach is replaced by more precise therapy decision-making, and if thus the mistreatment rate is reduced (resulting in a better image of allopathic medicine on a long-term basis), then the blind "cost effectiveness" approach of the currently still dominant "pharmacoeconomy" can be replaced by a differentiated cost/benefit picture. Within this picture, reduction of manufacturing costs for a therapeutic agent will not be the only decisive factor, but the reduction of all costs having to be incurred for successful therapy – including the transaction costs of the physician/patient therapy relationship. Reducing the transaction costs of the physician/patient contract relationship has so far been an almost unknown factor – a blind spot in the pharmacoeconomical debate. The market for pharmaceuticals and therapeutic aids has become increasingly intransparent – even for physicians. Hence, we need much more precise and highly differentiated product information. The manufacturers' product marketing in turn cannot be seen from a mere economical marketing viewpoint alone any longer; the manufacturers have to become aware of the fact that decisive therapy-related information has to be delivered here which will help reduce the overall macroeconomic costs by helping physicians administer more precise, to-the-point, agents and therapies.

Alternde Industrienationen im Vergleich. Eine Gegenüberstellung zwischen den USA, Deutschland und Japan
Comparing and contrasting the industrialized nations: the U.S., Germany and Japan

Von Burkhard Jacob
By Burkhard Jacob

Gesundheit ist unser kostbarstes Gut. Wir lernen sie erst richtig schätzen, wenn wir krank oder durch Unfall oder Alter eingeschränkt sind. Dabei wird längst nicht alles, was die Gesundheit fördert, auch vom Gesundheitssystem bezahlt, etwa wenn wir schwimmen gehen, Rad fahren oder einen Spaziergang machen, wenn wir unsere Zähne putzen, uns durch Kleidung vor schlechtem Wetter schützen oder uns in der Sauna erholen. Wir pflegen uns täglich, ohne krank zu sein. Pflege ist immer erforderlich, um gesund zu bleiben. Und Pflege bedeutet, sich eigenverantwortlich um sich selbst und andere zu kümmern, nicht nur bei Krankheit. Der Krankenstatus muss dagegen formell durch einen Arzt zuerkannt werden.

„Man kann ohne Behandlung durch Therapeuten wieder gesunden, aber man kann niemals ohne Ärzte krankgeschrieben werden", erklärt Johann Behrens, Direktor des Instituts für Gesundheits- und Pflegewissenschaft an der Universität Halle. „Kindheit ist keine Krankheit, Alter ist keine Krankheit ... eine Geburt ist keine Krankheit. Auch eine Behinderung ist keine Krankheit, sonst könnte ein Behinderter nicht krank und wieder gesund werden"[1], so Johann Behrens weiter.

Natürlich können auch Pflegebedürftige krank werden. Und angesichts des demografischen Wandels werden sich die alternden Gesellschaften Problemen bewusst, die insbesondere die Kranken- und Pflegeversicherungen auf eine Probe stellen. In den USA beträgt die Lebenserwartung von Frauen 81 Jahre, von Männern 75 Jahre. Frauen in Deutschland werden im Durchschnitt 82 Jahre alt, Männer 77 Jahre. In Japan liegt die Lebenserwartung von Frauen bei 86 Jahren, von Männern bei 79 Jahren.[2] Obgleich die Lebenserwartung in den industrialisierten Ländern zu den höchsten der Welt gehört, zeigt der Ländervergleich zwischen Deutschland, Japan und den USA, dass die Vorstellungen von Gesundheit, Krankheit und Alter recht unterschiedlich sein können. Dies ist kein Zufall, sondern das Ergebnis von kulturellen Entwicklungen.

Health is our most valuable asset. We tend to appreciate it most when we are ill or are limited by accident or age. For much of our lives good health comes not simply from paid healthcare and health promotion but from going for a swim, a bike ride or a walk, when we brush our teeth, wear protective clothing in bad weather or even relax in a sauna. We take care of ourselves everyday in order not to become sick. Care is always necessary in order to stay healthy. Such care means being responsible for oneself and others, and not only in sickness. Determination of sickness must be formally made by a doctor.

"One can recover without treatment from therapists, but one can never be determined ill without a doctor's examination," explains Johann Behrens, Director of the Institute of Health Science and Nursing Science at the University of Halle. "Childhood is not an illness, age is not an illness ... a birth is not an illness. Nor is a handicap an illness, otherwise a disabled person could not become ill and then recover,"[1] says Behrens further.

Of course people requiring care and assistance can become ill. In view of the demographic change, aging societies are realizing the problems which particularly strain health and long-term care or nursing insurances. In the USA, the life expectancy of women is 81 years and for men it is 75 years. Women in Germany have a life expectancy of 82 years on average while for men it is 77 years. In Japan, the life expectancy of women is 86 years and 79 years for men.[2] Although life expectancies in the industrialized countries are the highest in the world, a comparison between Germany, Japan and the USA shows that the images of health, illness and age can be rather different. This is not the result of chance, but the result of cultural developments.

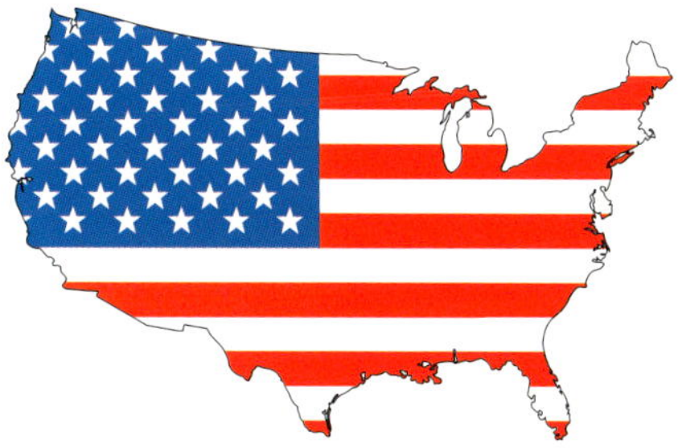

Das amerikanische Gesundheitssystem

Kaum ein Gesundheitssystem ist so facettenreich wie das US-amerikanische. Auf der einen Seite bringt es Spitzenmedizin, bahnbrechende Forschungsergebnisse und innovative Produkte hervor. Auf der anderen Seite sind rund 47 Millionen Amerikaner nicht oder nicht ausreichend krankenversichert. Und das Gesundheitssystem verschlingt riesige Summen. „Für viele Amerikaner ist die Krankenversicherung ein an den Arbeitsplatz gebundener Bonus oder eine Leistung aus Regierungsprogrammen“[3], erklärt David M. Cutler, Professor für Angewandte Volkswirtschaft an der Harvard University und Mitglied des Institute of Medicine. „Es scheint ironisch, dass die Nation mit den höchsten Pro-Kopf-Ausgaben für Gesundheit und dem technologisch fortschrittlichsten Gesundheitssystem der Welt nicht auch die gesündeste ist“, so Cutler.

Ein Grund mag in der besonders in Amerika ausgeprägten Idee von persönlicher Freiheit liegen, vom Recht jedes Einzelnen, seine eigene Lebensform zu wählen. Diese Form der Freiheit mündet allerdings nicht selten in einen „Individualismus der Selbstverwirklichung“[4]. So läuft das Ideal von Freiheit und Selbstbestimmung immer auch Gefahr, sich in einen konsumorientierten Individualismus mit bloßer Wahlfreiheit zu verwandeln. Eine ähnliche Diagnose stellt David Cutler von der Harvard University für das amerikanische Gesundheitswesen: „Das amerikanische Gesundheitswesen ist stark fragmentiert und folgt einem komplizierten Regelsystem. Eine Kombination aus privaten und staatlichen Bürokratien entscheidet, welcher Patient welche Behandlung erhält... Andererseits fühlen sich die Amerikaner mit einem fragmentierten System wohler als andere Nationen. Unser Volk misstraut allem, was vereinheitlicht ist – so auch ein System, das aus einem Topf die gesamte Krankenversicherung finanziert. Im Gesundheitswesen müssen wir diese kulturelle Vorliebe teuer bezahlen.“[5]

The American health system

No health system is as multi-faceted as the one found in the USA. On the one hand it produces top medicine, pioneering research results and innovative products. On the other hand, about 47 million Americans have either inadequate or simply no health insurance. Additionally, the healthcare system devours gigantic sums: “For many Americans, health insurance is a perk most often tied to their jobs, or it comes as a result of a government program such as Medicare or Medicaid,”[3] explains David M. Cutler, professor for applied national economy at Harvard University and member of the Institute of Medicine. He continues, “It is ironic that the United States, a nation that spends the most money per capita on healthcare and has the most technologically advanced medical system in the world, is not the healthiest society on earth.”

A reason for this may lie in the distinctively American idea of personal freedom, the right of individuals to choose their own lifestyle. However, this form of freedom leads not infrequently to an “individualism of self-realization.”[4] Thus the ideals of freedom and self-determination run the risk of morphing into an individualism oriented toward consumption with mere electoral freedom. A similar diagnosis comes from Harvard’s Cutler, “The American medical system is highly fragmented, with complicated rules and a combination of private and public bureaucracies that decide which patient can get what treatment ... On the other hand, Americans are more comfortable with fragmentation than other nationalities. As a nation, we are suspicious about a ‘single’ anything – like a single-payer system for health insurance. But in healthcare, our cultural preference comes at a high price.”[5]

In comparison to Germany and other European industrial nations such as France or Great Britain, the American healthcare system has been developed only relatively recently. When Franklin D. Roosevelt

Im Vergleich zu Deutschland und anderen europäischen Industrienationen wie Frankreich oder Großbritannien hat sich das amerikanische Gesundheitssystem erst relativ spät entwickelt. Als Franklin D. Roosevelt 1933 an die politische Macht kam, steckten die Vereinigten Staaten noch tief in der Wirtschaftskrise. Unter seiner Reformpolitik des „New Deal" wurde die staatliche Rente eingeführt und die Arbeitslosenversicherung ins Leben gerufen. Roosevelt förderte auch die Forschung nach einem Impfstoff gegen Polio. Mit 39 Jahren war er selbst an Kinderlähmung erkrankt und konnte seine Beine von da an kaum noch bewegen. Meistens war er auf den Rollstuhl angewiesen, stehen konnte er nur mit Hilfe von stählernen Beinschienen.

Die beiden Krankenversicherungen Medicaid und Medicare wurden erst 1965 im Rahmen des Social Security Act unter Präsident Lyndon B. Johnson eingeführt. Medicaid wird von den einzelnen Bundesstaaten organisiert und kann auch je nach Bundesstaat unterschiedliche Namen tragen. Zielgruppe sind Personenkreise mit geringem Einkommen, Kinder, ältere Menschen und Menschen mit Behinderungen. Jeder Bundesstaat hat ein eigenes Medicaid-Programm, das die Centers for Medicare and Medicaid Services (CMS) zentral überwachen. Die amerikanische Bundesregierung beteiligt sich an der Finanzierung.

Medicare ist für ältere und behinderte Mitbürger. Jeder Bürger ab dem Alter von 65 Jahren bzw. jeder als „behindert" anerkannte Bürger kann Medicare in Anspruch nehmen. „Obwohl öffentlich finanziert, handelt es sich bei Medicare streng genommen eher um eine Sozialversicherung. Ähnlich wie im deutschen System wird der Beitrag direkt vom Gehalt einbehalten und paritätisch von Arbeitgebern und Arbeitnehmern finanziert", erläutern Volker Eric Amelung und Katharina Janus von der Medizinischen Hochschule Hannover. Dabei bieten sich den Unternehmen erhebliche steuerliche Anreize, den Mitarbeitern Sozialleistungen wie eine Krankenversorgung anzubieten. „Die durchschnittlichen Kosten für eine Individualversicherung liegen bei 4.242 US-Dollar (ca. 3.200 Euro) pro Jahr, jene für eine Familienversicherung bei 11.483 US-Dollar (8.670 Euro). Dabei trägt der Arbeitgeber allerdings den Löwenanteil (durchschnittlich 84 Prozent bei Individualversicherungen und 73 Prozent bei Familienversicherungen). In den letzten Jahren haben sich immer mehr Unternehmen zu Einkaufsgenossenschaften zusammengeschlossen, um ihre Marktmacht zu steigern"[6], so Amelung und Janus.

Die US-amerikanischen Krankenversicherungen bieten je nach individueller Präferenz und Risikobereitschaft die unterschiedlichsten Policen und die verschiedensten Formen der Zuzahlung bzw. des Selbstbehalts an. Darüber hinaus hat sich die so genannte „Catastrophic Care Police" etabliert, bei der der Versicherte sämtliche Leistungen bis zu einer bestimmten Höhe selbst bezahlt. Neben der klassischen Krankenkasse können noch die „Health Maintenance Organizations" und „Preferred Provider Organisations" unterschieden werden. Im ersten Fall übernimmt die Organisation gegen einen Monats- oder Jahresbeitrag das finanzielle Risiko der Gesundheits-

became president in 1933, the United States was deep in the Great Depression. Under his reform policy of the New Deal, the new programs of social security and unemployment insurance were introduced. Roosevelt also promoted research for finding a polio vaccine. At the age of 39, he himself fell ill with polio and forever lost most of the use of his legs. Roosevelt depended on a wheelchair, as he could only stand with the help of steel leg braces.

Both Medicaid and Medicare only first became available in 1965 as part of the Social Security Act introduced under President Lyndon B. Johnson. Medicaid is managed individually by each state and is known by many different names from state to state. The target groups for Medicaid are those with low incomes, children, elderly, and people with disabilities. Each state has its own Medicaid program and all of these are overseen by the Centers for Medicare and Medicaid Services (CMS). The U.S. federal government contributes to the financing of Medicaid.

Medicare is a program for older and disabled citizens. Every citizen from the age of 65 or disabled citizens can take up Medicare. "Although financed with tax money, Medicare is strictly controlled and is not a social insurance. As in the German system, contributions are withheld directly from paychecks and equally charged to employers' taxes," explain Volker Eric Amelung and Katharina Janus of the Medical College of Hannover. To this end, employers are offered considerable tax incentives for providing healthcare benefits to their employees. "The average cost for individual insurance is about U.S. $4,242 (approx. €3,200) per year, for a family policy it is about U.S. $11,483 (€8,670). The employer, however, carries the lion's share (on average 84 % for an individual and 73 % for family policies). Over the last several years, more and more companies have gotten together to purchase their insurance plans in order to keep their prices down,"[6] according to Amelung and Janus.

The U.S. healthcare system offers the most different kinds of insurance policies and forms of payment, or rather deductibles, according to individual preferences and risk levels. In addition, there is also the so-called catastrophic insurance with which the insured person pays for all costs only up to a certain (catastrophic) value of his or her choosing. Besides the classic health insurance scheme, there is a distinction between Health Maintenance Organizations (HMO) and Preferred Provider Organizations (PPO). In the first case, the organization takes on the financial risk and assigns the healthcare based on a monthly contribution or a yearly subscription. In the second case, the insured person selects from the doctors, clinics and therapists of the organization. The insured person then pays a lower contribution of the costs.

Notwithstanding, the number of under- or uninsured people in the U.S. has been rising steadily. Despite the tax incentives, still not every employer offers health insurance to its employees. Younger target groups in particular often have no health insurance, with 61 % of people under 34 reportedly being uninsured. Clearly, Americans

VEREINIGTE STAATEN VON AMERIKA / USA
GESUNDHEITSDATEN HEALTH DATA

Ärzte pro 1.000 Einwohner
Physicians, density per 1,000 population
2,4

Krankenschwestern pro 1.000 Einwohner
Nurses, density per 1,000 population
9,4

Krankenhausbetten pro 1.000 Einwohner
Hospital beds, density per 1,000 population
2,7

Gesundheitsausgaben pro Kopf in 2007
Health expenditure per capita, 2007
7,290 $

Lebenserwartung von Frauen bei der Geburt
Life expectancy of women at birth
81 Jahre

Lebenserwartung von Männern bei der Geburt
Life expectancy of men at birth
75 Jahre

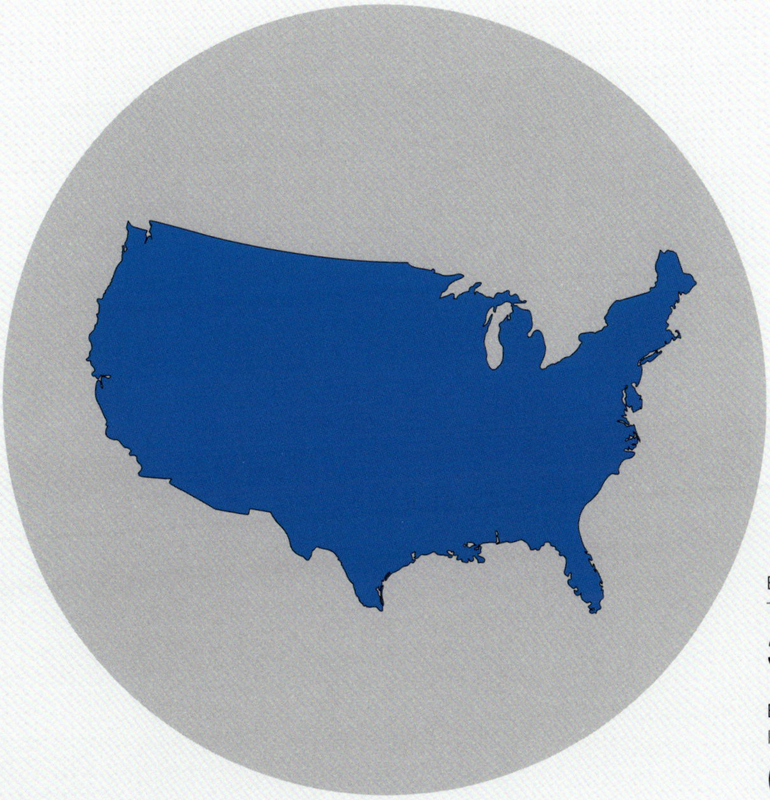

Bevölkerung in Millionen
Total population in millions
304,5

Bevölkerung Wachstumsrate pro Jahr
Population growth rate
0,6%

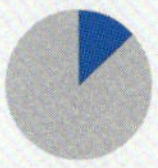

Gesundheitsausgaben gemessen am Bruttoinlandsprodukt BIP in 2000
Health expenditure as share of gross domestic product (GDP), 2000
13,2%

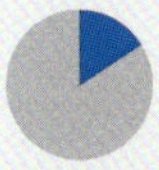

Gesundheitsausgaben gemessen am Bruttoinlandsprodukt BIP in 2007
Health expenditure as share of gross domestic product (GDP), 2007
16 %

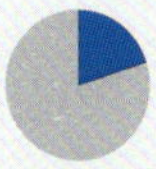

Anteil der Bevölkerung unter 15 Jahren
Percentage of population aged 15 and younger
20%

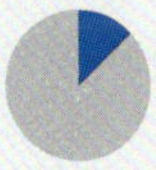

Anteil der Bevölkerung über 65 Jahre
Percentage of population aged 65 and older
13%

Quellen / Sources: OECD Health Data 2008, OECD Factbook 2009, German Foundation for World Population 2009

versorgung. Im zweiten Fall verpflichtet sich der Versicherte, nur die Ärzte, Kliniken und Therapeuten der Organisation in Anspruch zu nehmen. Er zahlt im Gegenzug geringere Beiträge.

Gleichwohl ist die Zahl der Nicht- oder Unterversicherten in den USA stetig gestiegen, denn trotz Steueranreizen bietet nicht jeder Arbeitgeber eine Krankenversicherung als Bestandteil des Arbeitsvertrages an. Insbesondere bei den jüngeren Zielgruppen fehlt häufig ein Krankenversicherungsschutz. 61 Prozent ohne Versicherung sind jünger als 34 Jahre. Offenbar wollen die Amerikaner nicht noch mehr Geld für ein System zahlen, das sie selbst für marode halten. So ist das Thema Gesundheit in den USA längst zu einer Kostendebatte geworden.

Als Volkswirtschaftler sieht David Cutler von der Harvard University die Lösung vor allem in der Rationalisierung des Gesundheitswesens: „Wenn wir die Kranken davor bewahren können, noch kränker zu werden, können wir Geld sparen." So schlägt er ein Bündel unterschiedlicher Strategien vor, die unter anderem die Vernetzung des Gesundheitswesens und die Überprüfung der Wirksamkeit von Medikamenten umfasst und mehr Zeit und Energie auf die Prävention verwendet. Trotz der berechtigten Forderungen wird man angesichts der Finanzkrise und Rezession von der Regierung Obama keine medizinischen Wunder erwarten dürfen. Gleichwohl hat Obama mit der Reform des Gesundheitswesens in den USA eines der schwierigsten innenpolitischen Projekte in Angriff genommen. Am 15. Juni 2009 stellte er in seiner Rede auf der Konferenz der „American Medical Association" in Chicago fest: „Presidents have called for health care reform for nearly a century. Teddy Roosevelt called for it. Harry Truman called for it. Richard Nixon called for it. Jimmy Carter called for it. Bill Clinton called for it. But while significant individual reforms have been made – such as Medicare, Medicaid, and the children's health insurance program – efforts at comprehensive reform that covers everyone and brings down costs have largely failed ... The question now is, how do we finish the job? How do we permanently bring down costs and make quality, affordable health care available to every American? ... My view is that health care reform should be guided by a simple principle: fix what's broken and build on what works."[7]

Das deutsche und japanische Gesundheitssystem

Es mag paradox klingen, aber die Erfolge des medizinischen Fortschritts haben nicht nur in den USA dazu beigetragen, dass das Gesundheitssystem unter den eigenen Kosten zu kollabieren droht. Auch in Deutschland und Japan hat die Kostendebatte zu einer intensiven Auseinandersetzung mit Grundlagen des Wohlfahrtsstaates geführt. Der Blick in die Geschichte zeigt für beide Länder zunächst ähnliche Ausgangsbedingungen am Ende des 19. und zu Beginn des 20. Jahrhunderts. Die Entwicklung des Gesundheitssystems verläuft dann aber unterschiedlich, was nicht zuletzt auf die gesellschaftlichen und kulturellen Unterschiede zurückzuführen ist.

do not want to pay even more money for a system which they themselves consider to be ailing. Thus, the subject of healthcare in America has become a critical cost debate.

As an economist, Cutler sees the solution above all in the rationalization of the healthcare service: "If we can keep the sick from getting sicker, we should be able to save money." Thus, he suggests a bundle of different strategies which address an interlinking of the healthcare service, an examination of the effectiveness of drugs and focuses more time and energy on prevention. In spite of these reasonable demands, and in light of the current economic crisis, no one can truly expect miracle reforms to the healthcare system from the Obama administration. Nevertheless, President Obama has started trying to tackle one of the most difficult domestic projects and addressed healthcare reform at a conference of the American Medical Association in Chicago on 15 June 2009. "Presidents have called for healthcare reform for nearly a century. Teddy Roosevelt called for it. Harry Truman called for it. Richard Nixon called for it. Jimmy Carter called for it. Bill Clinton called for it. But while significant individual reforms have been made – such as Medicare, Medicaid, and the children's health insurance program – efforts at comprehensive reform that covers everyone and brings down costs have largely failed ... The question now is, how do we finish the job? How do we permanently bring down costs and make quality, affordable health care available to every American? ... My view is that health care reform should be guided by a simple principle: fix what's broken and build on what works."[7]

The German and Japanese healthcare systems

It may sound paradoxical, but the success of medical advancements has not only threatened the collapse of the U.S. healthcare system under its own weight, but rising costs in Germany and Japan have also led to intense discussions about the foundations of those welfare states. A look at their history shows similar starting points for both countries at the end of the 19th and the beginning of the 20th centuries. However, one can see the different development of the healthcare systems, which can best be attributed to their social and cultural differences.

In both Germany and Japan the industrialization revolution, which had kicked off in England, got going a little bit late. Despite this, the two late-comers caught up quickly, forcing a rather abrupt discussion in the latter half of the 19th century about what to do with the newly created working class. In Germany, the welfare policies were decided by Chancellor Otto von Bismarck until 1890. Bismarck's welfare legislation was a breakthrough in the concept of insurance and of the question of welfare in Germany.

In 1883, Bismarck introduced legal health insurance which was predominately aimed at the work force. "My thought was to win over the working classes, or rather to fascinate them; to look at the state

In Deutschland und Japan setzte die von England ausgehende Industrialisierung etwas verspätet ein. Gleichwohl holten die beiden Nachzügler schnell auf, sodass sich insbesondere die soziale Frage mit Blick auf die neu entstandene Arbeiterklasse in der zweiten Hälfte des 19. Jahrhunderts umso heftiger stellte. In Deutschland wurde die Sozialpolitik bis 1890 entscheidend durch Reichskanzler Otto von Bismarck geprägt. Seine Sozialgesetzgebung stellte einen Durchbruch für den Versicherungsgedanken und die soziale Frage in Deutschland dar.

1883 führte Bismarck die gesetzliche Krankenversicherung ein, die überwiegend auf die Arbeiterschaft ausgerichtet war. „Mein Gedanke war, die arbeitenden Klassen zu gewinnen, oder soll ich sagen zu bestechen, den Staat als soziale Einrichtung anzusehen, die ihretwegen besteht und für ihr Wohl sorgen möchte“[8], so Bismarck.

Über Bismarcks Motive ist viel geschrieben worden und man mag trefflich darüber streiten, gleichwohl rief er 1883 die gesetzliche Krankenversicherung ins Leben. 1884 folgte die Unfallversicherung und 1889 die gesetzliche Rentenversicherung, die ursprünglich als Invaliditäts- und Altersversicherung konzipiert war. In Deutschland entstand ein überwiegend beitragsfinanziertes System, das in der sozialen Frage eine zukunftsweisende Rolle sowohl für Europa als auch gegenüber den USA und Japan einnahm. 1928, während der Weimarer Republik, wurde schließlich die Arbeitslosenversicherung in Deutschland eingeführt, um die Kassen der Kommunen zu entlasten, allerdings wurden die Leistungen im Wohlfahrtssektor unter dem Eindruck der Weltwirtschaftskrise durch Inflation und durch die spätere Entfremdung der Mittel zur Rüstungsfinanzierung immer weiter ausgehöhlt.

Unter dem Eindruck der Industrialisierung hatten sich auch in Japan die alten ständischen Strukturen und Milieus, bestehend aus Samurai, Kaufleuten, Handwerkern und Bauern, aufgelöst, gleichwohl musste sich eine neue nationalstaatliche Gesellschaft auch hier erst herausbilden. Anders als in den USA basiert die Idee des Wohlfahrtsstaates in Japan aber nicht auf Individualismus. Und anders als in Deutschland, wo das mit der Industrialisierung erzeugte Solidaritätsproblem durch den Sozialstaat bismarckscher Prägung beantwortet wurde – und damit auch die christliche Religion aus ihrer Rolle der Solidarität langsam in den Hintergrund gedrängt wurde – stärkte man in Japan die Rolle der Familie. „Hier wurde geradezu die Übertragbarkeit der Familienmetapher auf die Gesamtgesellschaft unterstrichen und damit das Zusammengehörigkeitsgefühl angeregt. Es war immer das japanische Haushaltskonzept (ie = Haus), das bis zur heutigen Zeit zum zentralen Bereich der kulturellen Identität gezählt wird. Auffällig ist dabei, dass traditionelle Elemente der sozialen Bindung aus den Bereichen der Familie, Religion und Dorfgemeinschaft für die Konstituierung der gesamtgesellschaftlichen Solidarität aktiviert und eingesetzt wurden. Auf diese Weise gelang dem japanischen Nationalstaat die Konstituierung einer Form von Solidarität, obgleich dies später zu einem kulturfundamentalistischen Militarismus führte. Aber auch nach dem Zweiten Weltkrieg blieb die Idee

as a welfare tool which exists for them and that would like to provide for their welfare,"[8] said Bismarck.

A lot has been written about Bismarck's motives and one could argue vigorously about it, but in any case, in 1883 he brought legal health insurance to life. In 1884, accident insurance followed and then in 1889 the legal pension scheme, which was originally conceived of as disability and old age insurance, was introduced. A system of contribution-financed insurance began in Germany and sparked a social welfare trend across Europe and even reaching as far as Japan and the U.S. In 1928, during the Weimar Republic, unemployment insurance was introduced in Germany to relieve the coffers of the local authority districts. Of course, the advancements this made were hollowed out under the pressures of the world-wide economic crisis of the time through high inflation and later by the continuing appropriation of funds for the military.

In Japan, the old corporate structures and environments, consisting of samurai, businesspeople, craftsmen and farmers, also started to break apart requiring Japan to reinvent its nation-state for a new kind of society. Unlike in the USA, Japan's idea of the social welfare state is not based on individualism. And unlike in Germany where the solidarity problem stemming from industrialization was answered by Bismarck's brand of the social welfare state – and the role Christianity played in slowly pushing solidarity to the background – Japan focused on strengthening the role of the family. "Here the transferability of the family metaphor onto the whole society was strongly emphasized and with it was inspired the feeling of belonging. It has always been the Japanese concept of household (ie = house) which has been central to the concept of cultural identity. Not insignificantly, traditional elements of social belonging from the areas of family, religion and village life were activated and adopted in order to establish solidarity throughout the entire society. In this manner, the Japanese nation-state managed to establish a form of solidarity, although this later led to a culturally fundamentalist militarism. Still, though, after World War II the idea remained that Japanese society is a homogeneous society in which the family structure and familial relationships prevail together,"[9] as described by Shingo Shimada and Christian Tagsold. Both teach in the philosophy department of the Heinrich Heine University in Düsseldorf. The supporting role of the family can be traced back to the Japanese religions of Shintō and Buddhism, as well as to the political philosophy influenced by Chinese Confucianism.

According to the teachings of Confucius, the basic relationships were between ruler and subject, parent and child, husband and wife, older and younger siblings, and between casual friends. The family, therefore, had on the one hand an outstanding position; on the other hand, there was a very strong hierarchy to the relationships. The neo-Confucianist image of help corresponded not to a familial bond and the Christian idea of charity. Buddhist institutions and temples could also not develop large welfare institutions as the churches in Germany had successfully done with charity and deacons. Instead,

erhalten, dass die japanische Gesellschaft eine homogene Gesellschaft bilde, in der die familienähnliche Beziehung untereinander vorherrschend sei"[9], wie Shingo Shimada und Christian Tagsold die japanische Situation beschreiben. Beide lehren an der Philosophischen Fakultät der Heinrich-Heine-Universität Düsseldorf. Die tragende Rolle der Familie lässt sich bis in die japanischen Religionen von Shintō und Buddhismus sowie bis in die durch den chinesischen Konfuzianismus beeinflusste politische Philosophie zurückverfolgen.

Der Lehre des Konfuzius folgend waren die grundlegenden Beziehungen zwischen Herrscher und Untertan, Eltern und Kindern, Ehemann und Ehefrau, älteren und jüngeren Geschwistern sowie zwischen Freunden von grundlegender Bedeutung. Damit hatte die Familie einerseits eine herausgehobene Stellung, andererseits waren die Beziehungen sehr stark hierarchisiert. Die neokonfuzianische Vorstellung von Hilfe entsprach also nicht einer familiären Bindung und Hilfe im Sinne christlicher Nächstenliebe. Die buddhistischen Institutionen und Tempel konnten auch keine großen Wohlfahrtsinstitutionen ausbilden, wie dies die Kirchen in Deutschland mit Caritas und Diakonie erfolgreich getan hatten. Stattdessen wurde 1868 der Shintoismus (shintō = Weg der Götter) zur Staatsreligion, der auf die Person des Tennō (tennō = Himmlischer Herrscher) hin zentriert wurde. Somit wurde im Zuge der Öffnung Japans zum Westen und mit der Industrialisierung auch die über 250 Jahre währende Herrschaft der Tokugawa-Shōgun-Dynastie, die dem so genannten Kriegeradel (Bushi oder Samurai) angehörten und eine Einteilung in die vier erblichen Stände geschaffen hatten, schrittweise aufgelöst.

Folgerichtig ging die Einrichtung erster sozialer Sicherungssysteme einseitig vom Staat aus und erstreckte sich zunächst nur auf Staatsdiener. Kurze Zeit nach der Erklärung des Meiji-Tennō (1868), alle politische Macht in sich zu vereinen, wurden Alters- und Invalidenpensionen für Soldaten der Armee beschlossen. 1875 folgten Soldaten der Marine, 1882 die Polizei und 1884 schließlich auch die Zivilbeamten.

Somit stellte sich die Frage der sozialen Absicherung der Arbeiter zunächst nicht, obgleich auch in Japan die Industrialisierung zügig voranschritt und soziale Fragen entstanden. Shingo Shimada und Christian Tagsold von der Heinrich-Heine-Universität Düsseldorf dazu: „Eine Risikovorsorge für Arbeiter begann erst durch die Einrichtung betrieblicher Firmenkassen zur Jahrhundertwende. Im Mittelpunkt der staatlichen Wohlfahrtskonzeption stand der Verweis auf die im ie (Haus) lebende Familie, während eine Einrichtung nach bismarckschem Vorbild nicht verfolgt wurde."[10]

1926 begann mit Hirohitos Inthronisierung die Shōwa-Zeit (shouwa jidai = Ära des erleuchteten Friedens). 1927 wurde unter seiner Führung das Gesetz zur Einführung einer allgemeinen Krankenversicherung vom japanischen Parlament verabschiedet. Allerdings wurde auch Japan von der Weltwirtschaftskrise getroffen. Unter dem Eindruck der Krise zog vor allem die Armee das Gesetz des Han-

in 1868 Shintō (shintō = way of the gods) became the official state religion, whereby the focus was on the Tennō (tennō = heavenly emperor). Throughout the process of Japan's opening itself to the West and its industrialization, the more than 250 year old Tokugawa Shōgun-Dynasty, to which the so-called warrior's nobility (bushi or samurai) belonged, gradually disintegrated resulting in the division of the four inherited states.

The features of the first social security systems consistently emanated unilaterally from the state and at first only applied to public servants. A short time after the restoration of power to the Meiji-Tennō (1868), whereby all political power was restored to the emperor, it was decided that age and disability pensions would be granted for soldiers in the army; in 1875 navy soldiers followed, in 1882 it was granted to the police and finally in 1884 it was offered to civil servants.

So the question of workers' social security did not come up at first, although industrialization was progressing quickly in Japan and social questions did arise. Shingo Shimada and Christian Tagsold of the Heinrich Heine University in Düsseldorf explain it this way: "The thought of risk prevention for workers only first began with the establishment of internal company coffers at the turn of the century. At the center of the state welfare concept stood the reference to the ie (house) living family, whereas an arrangement after Bismarck's model would not be pursued."[10]

The Shōwa period (shouwa jidai = period of enlightened peace) began in 1926 with Hirohito's coronation. In 1927 under the leadership of Hirohito, the Japanese parliament adopted a general health insurance plan into law. Certainly Japan was also gripped by the worldwide economic crisis of the time. Under pressure from the crisis, the military more than anyone else used the law of business to its own advantage and moved the Japanese economy more and more toward the needs of the military and a wartime economy. In 1941 and 1942, in the midst of the war, Japan succeeded even in dismissing the workers' pension scheme law, which already would be revised in 1944 and renamed as the welfare pension scheme. The law formally oriented itself to the German social security system. The contributions were made half by the employees and half by the employers.

Through their unconditional surrender to the allies Germany and Japan were reshaped after World War II. In Germany, the social state principle is led by Article 20 of German Constitutional Law. In Japan, the new constitution sees the Tennō only as a ceremonial symbol, no longer as an actual reference point of the state religion and the political system. "Accordingly, marriage was addressed in Article 24, which established the equality of both partners and thus annulled the ie system as a basis of family law."[11]

After losing the war, both Germany and Japan had to face regaining their sovereignty and the reconstruction of their countries above all

DEUTSCHLAND / GERMANY
GESUNDHEITSDATEN HEALTH DATA

Ärzte pro 1.000 Einwohner
Physicians, density per 1,000 population
3,5

Krankenschwestern pro 1.000 Einwohner
Nurses, density per 1,000 population
9,8

Lebenserwartung von Frauen bei der Geburt
Life expectancy of women at birth
82 Jahre

Krankenhausbetten pro 1.000 Einwohner
Hospital beds, density per 1,000 population
6,2

Lebenserwartung von Männern bei der Geburt
Life expectancy of men at birth
77 Jahre

Gesundheitsausgaben pro Kopf in 2007
Health expenditure per capita, 2007
3,588 $

Bevölkerung in Millionen
Total population in millions
82,2

Bevölkerung Wachstumsrate pro Jahr
Population growth rate
-0,2%

Gesundheitsausgaben gemessen am Bruttoinlandsprodukt BIP in 2000
Health expenditure as share of gross domestic product (GDP), 2000
10,3%

Anteil der Bevölkerung unter 15 Jahren
Percentage of population aged 15 and younger
14%

Gesundheitsausgaben gemessen am Bruttoinlandsprodukt BIP in 2007
Health expenditure as share of gross domestic product (GDP), 2007
10,4%

Anteil der Bevölkerung über 65 Jahre
Percentage of population aged 65 and older
20%

Quellen / Sources: OECD Health Data 2008, OECD Factbook 2009, German Foundation for World Population 2009

delns an sich und stellte die japanische Wirtschaft mehr und mehr auf die Bedürfnisse der Militärs und der Kriegswirtschaft um. 1941/42, inmitten des Krieges, gelang es indes, das Arbeiterrentenversicherungsgesetz zu verabschieden, das bereits 1944 revidiert und in die Wohlfahrtsrentenversicherung umbenannt wurde. Formal orientierte sich das Gesetz an den deutschen Sozialversicherungen, die Beiträge wurden also je zur Hälfte von den Arbeitnehmern und Arbeitgebern geleistet.

Mit der bedingungslosen Kapitulation und der Besatzung durch die Alliierten wurden Deutschland und Japan nach dem Zweiten Weltkrieg umgestaltet. In Deutschland leitet sich das Sozialstaatsprinzip seither aus Artikel 20 des Grundgesetzes ab. In Japan sieht die neue Verfassung den Tennõ nur noch als zeremonielles Symbol, nicht mehr als realen Bezugspunkt der Staatsreligion und des politischen Systems: „Entsprechend wurde auch der Artikel 24 zur Ehe formuliert, der die Gleichstellung beider Ehepartner festschrieb und so das ie-System als Grundlage des Familienrechts aushebelte."[11]

Nach dem verlorenen Krieg standen sowohl für Deutschland als auch für Japan die Wiedererlangung der Souveränität und der Wiederaufbau des Landes im Vordergrund. Rückblickend betrachtet erlebten beide Länder die Zeit des wirtschaftlichen Aufschwungs als Wirtschaftswunder. Prägend für die Zeit in Deutschland war Ludwig Erhards Buch „Wohlstand für alle", mit dem er 1957 als Bundeswirtschaftsminister seine Vorstellungen von Marktwirtschaft publizierte.

In Japan verkündete 1973 der damalige japanische Premierminister Tanaka Kakuei gar das „Wohlfahrtsgründungsjahr". Die Ölkrise verhinderte allerdings eine drastische Erhöhung der Sozialausgaben. Unter dem neuen Premierminister Miki Takeo wurde dann die übergreifende Agenda einer „Japanese style welfare society" verkündet, die Eigenheime für alle Japaner, einen sozialen Mindestschutz und einen sicheren Lebensabend für Ältere plante. Letztlich sollte die Agenda aber nur als Schlagwort überleben, da insbesondere die Pflege Älterer auf die Familien und damit letztlich auf die Frauen abgeschoben wurde. Tatsächlich blieben die japanischen Aufwendungen für Sozialleistungen bis in die 1990er Jahre deutlich hinter denen in Europa zurück. Gemessen an der Sozialleistungsquote, also dem Anteil der Sozialleistungen am Bruttoinlandsprodukt, lag Japan mit etwa 12,5 Prozent deutlich hinter Deutschland, das 1990 eine Quote von 23,5 Prozent aufweisen konnte.

1995 wurde in Deutschland die Pflegeversicherung eingeführt. Sie steht in der langen Tradition der gesetzlichen Sozialversicherungen, die unter Bismarck ins Leben gerufen wurden. Vor Einführung der Pflegeversicherung wurden alte Menschen, die die Kosten der Pflege nicht mehr selbst aufbringen konnten, in der Regel zu Sozialfällen und hatten damit Anspruch auf Versorgung aus den Kassen der Sozialhilfe. Die Kosten mussten von den Kommunen getragen werden.

else. Looking back, both countries experienced this time of economic recovery as an economic miracle. Formative in Germany at the time was Minister of Economic Affairs, Ludwig Erhard's 1957 book, Prosperity for Everyone, in which he published his ideas of a market economy. Japan's Prime Minister, Tanaka Kakuei, even announced in 1973 the Inaugural Welfare Budget. The oil crisis certainly prevented a drastic rise in welfare expenses. Under the new Prime Minister Miki Takeo, an overall agenda for a Japanese-style welfare society was announced; one which planned for home ownership for all, a minimum social protection and security in old age. In the end, however, the agenda only survived as a catchword, since particularly the care of the elderly was pushed on to the families, and ultimately, on to the women. Actually, Japanese expenditures for social services lagged well behind those in Europe until the 1990s. Measured in terms of social services quotas, the portion of Japan's social services paid for as a part of the country's gross domestic product was 12.5 %, clearly behind Germany, which itself measured at 23.5 % in 1990.

In 1995, long-term care insurance was introduced in Germany. It counts among a long tradition of legal welfare insurances first brought to life under Bismarck. Before the introduction of long-term care insurance, elderly people who could not bear the costs of such care themselves, as a rule became social cases and thus claimed the costs from the state welfare budget. The costs then had to be carried by the local authority districts. In 2000, Japan followed the German model and introduced its own version of long-term care insurance. While in Germany a need for action arose from the situation between the nursing dependent and the local authority districts, in Japan it was the situation with the families for whom the care-giving was no longer reasonable. In everyday life this fell to the women, often to the daughters and daughters-in-law who were already loaded down by household chores and their children's educations. "A way out from the nursing hell for the families was either to put the ailing person in a home or hospital,"[12] say Shimada and Tagsold about the situation in Japan. Hospitalization led in particular to new social problems. A lack of availability led to long waiting periods. Space allocation was managed centrally under municipal supervision. Other options were not available. While the costs for placing someone in a home or hospital were still manageable for families, they destroyed the budgets of the health insurance schemes.

Upon closer consideration of the healthcare systems and the welfare state programs, there are significant differences between Germany, Japan and the USA. A look at long-term care insurances makes this especially clear. In the USA, there is no such legal pay-as-you-go long-term care insurance. There is instead a capital-covered, private insurance for financial security and safeguarding in old age, faithful to the motto: more individual responsibility, less solidarity. In Germany, this kind of care was anchored early on to the municipal level and today it is a required part of the long-term and nursing care budgets which are connected at regional and national levels to the

JAPAN / JAPAN

GESUNDHEITSDATEN HEALTH DATA

Lebenserwartung von Frauen bei der Geburt
Life expectancy of women at birth
86 Jahre

Lebenserwartung von Männern bei der Geburt
Life expectancy of men at birth
79 Jahre

Ärzte pro 1.000 Einwohner
Physicians, density per 1,000 population
2,1

Krankenschwestern pro 1.000 Einwohner
Nurses, density per 1,000 population
9,3

Krankenhausbetten pro 1.000 Einwohner
Hospital beds, density per 1,000 population
8,2

Gesundheitsausgaben pro Kopf in 2006
Health expenditure per capita, 2006
2,581 $

Bevölkerung in Millionen
Total population in millions
127,7

Bevölkerung Wachstumsrate pro Jahr
Population growth rate
0,0%

Gesundheitsausgaben gemessen am Bruttoinlandsprodukt BIP in 2000
Health expenditure as share of gross domestic product (GDP), 2000
7,7%

Gesundheitsausgaben gemessen am Bruttoinlandsprodukt BIP in 2006
Health expenditure as share of gross domestic product (GDP), 2006
8,1%

Anteil der Bevölkerung unter 15 Jahren
Percentage of population aged 15 and younger
13%

Anteil der Bevölkerung über 65 Jahre
Percentage of population aged 65 and older
23%

Quellen / Sources: OECD Health Data 2008, OECD Factbook 2009, German Foundation for World Population 2009

Im Jahr 2000 folgten die Japaner dem deutschen Vorbild und führten ihre Version der Pflegeversicherung ein. Während sich in Deutschland aus der Situation der Pflegebedürftigen und der Kommunen ein konkreter Handlungsbedarf ergab, war es in Japan die Situation der Familien, denen die Pflege nicht mehr zumutbar war. Im Alltag war dies vor allem die Aufgabe der Frauen, häufig der Töchter und Schwiegertöchter, die durch den Haushalt und die Erziehung der Kinder gleich mehrfach belastet waren. „Ein Ausweg aus der Pflegehölle für die Familien war die Heimunterbringung oder die Einweisung in ein Krankenhaus"[12], beschreiben Shimada und Tagsold die Situation in Japan. Aber insbesondere die Hospitalisierung führte zu neuen sozialen Problemen. Der Mangel an Heimplätzen führte zu langen Wartezeiten. Die Zuweisung erfolgte zentral unter kommunaler Aufsicht. Eine Wahlmöglichkeit war nicht gegeben. Während die Kosten für Unterbringung in Heimen und Krankenhäusern für die Familien noch überschaubar waren, sprengten sie die Budgets der Krankenkassen.

Bei genauerer Betrachtung des Gesundheitssystems und der sozialstaatlichen Programme gibt es also signifikante Unterschiede zwischen den USA, Deutschland und Japan. Der Blick auf die Pflegeversicherung macht dies besonders deutlich. In den USA gibt es diese Form der umlagefinanzierten gesetzlichen Pflegeversicherung nicht. Hier dient eine kapitalgedeckte private Versicherung der Vorsorge und Absicherung im Alter, getreu dem Motto: mehr Eigenverantwortung, weniger Solidarität. In Deutschland war die Pflege früher auf kommunaler Ebene verankert, heute obliegt sie den Pflegekassen, die auf regionaler und nationaler Ebene an die gesetzlichen Krankenkassen angeschlossen sind. Dabei wird die Feststellung der Pflegebedürftigkeit in einer von drei Stufen durch einen von der Pflegekasse bestellten Arzt festgestellt. Und dabei gilt: Nur wenn die Leistungspflicht nicht die Krankenkasse trifft, liegt die Erstattung bei der Pflegeversicherung. Die Versicherten erhalten dann Sachleistungen oder Pflegegeld, die sich nach dem Grad der Pflegebedürftigkeit richten. Die Krankenkasse ist gegenüber der Pflegeversicherung zunächst also primär, d. h., die Erstattung von Hilfsmitteln liegt in erster Linie bei der Krankenversicherung. In Japan wurde dagegen die Rolle der Kommunen gestärkt. Sie wurden zum Versicherungsträger und entscheiden seither über die Beitragssätze. Anders als in Deutschland, wo die Pflegeversicherung altersunabhängig ist, bleibt das japanische Modell auf eine reine Alterspflege beschränkt. Entsprechend sind nur über 40-Jährige beitragspflichtig. Zudem hat man in Japan die Position des „Care-Managers" geschaffen. Im Unterschied zu Deutschland, wo der Pflegedienst mit den Bedürftigen und den Angehörigen die Art der Pflege direkt abstimmt, koordiniert der „Care-Manager" in Japan die Art der Pflegedienstleistung. Er prüft die Bedürfnisse und hält den Kontakt zur Familie des Bedürftigen, wählt Anbieter unter den Pflegedienstleistern aus und erstellt einen verbindlichen Pflegeplan. Anders als in Deutschland werden in Japan fünf Pflegestufen und eine Vorstufe unterschieden. Es können allerdings nur Sachleistungen bezogen werden.

compulsory health insurance fund. As such, the determination of long-term care or nursing need is ascertained in one of three steps by a doctor assigned by the public healthcare system. The condition is: only if the liability duty does not fall on the health insurance fund is the reimbursement the responsibility of the long-term care insurance. The insured party receives an allowance in kind or a nursing allowance, the amount of which is determined by the degree of care needed. Compared with the long-term care insurance, the health insurance fund is primary, i.e. the reimbursement of aids lies, primarily, with the health insurance coverage. In Japan, the role of the local authority districts was strengthened against this. They became the underwriters and took on the responsibility of deciding the contribution rates. Unlike in Germany where long-term care insurance is provided irrespective of age, the Japanese model remains restricted to old age care. Accordingly, only people over 40 years old may make contributions to such insurance. In addition, the position of care manager has been created. As opposed to Germany where the kind of care provided is discussed with and approved by family members of the patient, the care manager in Japan coordinates the kind of service to be provided. The care manager checks the needs and maintains contact with the patient's family, selects the care provider among those available and develops a binding nursing care plan. Unlike in Germany, Japan has five care levels as well as a preliminary stage. Only allowances in kind, however, can be obtained.

The retroactive and hastily introduced structures to long-term care insurance cannot hide the fact, however, that the Japanese healthcare system is in crisis, just as Seiritsu Ogura, professor of economics at the Hosei University in Tokyo and editor of the Japanese Journal of Health Economics and Policy, ascertains:

"Unlike past crises, this one is a very quiet one, as it is coming from disappearing essential services. On weekends, we find emergency rooms (ERs) of any major hospitals practically deserted, as patients are turned down before – and after – their arrivals. The hospitals don't have enough staff to operate the ERs, and, even if they did, they couldn't afford to take care of the patient volume. In rural communities, local governments are closing their public hospitals in large numbers due to lack of funds. We have serious shortages of physicians, nurses, emergency facilities, pediatric hospitals, obstetric facilities, etc. We have observed a series of bad decisions by the Ministry of Health, Labor and Welfare during the past decade that should have contributed to this crisis, but they are more results than causes. We have a shortage of funds because we have not infused enough money into our health insurance system to care for the growing population of elderly. In the past few years, no one wanted to talk about tax increases for fear of losing the next election. The government has been piling up huge national debt for almost two decades, which has reached almost twice the size of the GDP. The implicit debt in our public pension program is almost the same size, too."[13]

Die nachträglich rasch eingeführten Strukturen zur Pflegeversicherung können aber nicht darüber hinwegtäuschen, dass sich das japanische Gesundheitssystem in einer Krise befindet, wie Seiritsu Ogura, Professor für Wirtschaft an der Hosei-Universität in Tokio und Redakteur des Japanese Journal of Health Economics and Policy, feststellt:

„Im Gegensatz zu vergangenen Krisen läuft diese jedoch in aller Stille ab, denn sie ist darauf zurückzuführen, dass wesentliche Dienstleistungen einfach verschwinden. An Wochenenden herrscht in der Notaufnahme aller großen Krankenhäuser gähnende Leere, weil Patienten entweder schon vor oder gleich nach ihrer Ankunft abgewiesen werden. Die Krankenhäuser haben nicht genug Personal für die Notaufnahme – und selbst wenn sie es hätten, könnten sie sich die Versorgung der vielen Patienten nicht leisten. In ländlichen Gemeinden werden zahlreiche öffentliche Krankenhäuser von den Kommunalverwaltungen aus Geldmangel geschlossen. Wir leiden an einem gravierenden Mangel an Ärzten, Krankenschwestern, Unfallstationen, Kinderkliniken, Geburtsstationen ... In den letzten zehn Jahren haben wir beobachtet, wie eine ganze Reihe schlechter Entscheidungen des Ministeriums für Gesundheit, Arbeit und Soziales zu dieser Situation beigetragen hat. Aber sie sind eher das Ergebnis der Krise als deren Ursache. Unsere Mittel sind knapp, weil wir nicht genug Geld in unser Krankenversicherungssystem gesteckt haben, um für den wachsenden Anteil Älterer in der Bevölkerung zu sorgen. In den letzten Jahren wollte niemand über Steuererhöhungen reden – aus Angst, die nächste Wahl zu verlieren. Der Staat hat in den letzten beiden Jahrzehnten einen riesigen Schuldenberg angehäuft, der inzwischen fast doppelt so groß wie das Bruttoinlandsprodukt ist. Das faktische Defizit unserer öffentlichen Rentenkasse ist ähnlich groß.“[13]

Der Vergleich der sozialstaatlichen Strukturen zwischen Japan, Deutschland und den USA, insbesondere die Gegenüberstellung der Gesundheitssysteme, macht aber auch deutlich, dass die Frage der Solidarität in erster Linie eine nationale Angelegenheit bleibt. Trotz menschlicher Verbundenheit über unterschiedliche Kulturkreise hinweg verbleibt diese Form der Solidarität innerhalb einer Gruppe von Gleichgesinnten oder innerhalb eines sozialstaatlichen Systems. Dabei dürften die Probleme der unterschiedlichen Länder ähnlich sein. Leider sind sie sich im Ergebnis schon so ähnlich, dass sich die Frage der Solidarität vielfach auf ein reines Kostenproblem reduziert hat. Dieses Problem gilt es zu lösen und die Menschen darüber hinaus für die Idee der Solidarität zu sensibilisieren. Solidarität gehört zum Erbe einer Kultur der Menschlichkeit, des Mitgefühls und der Verantwortung. Auch wenn diese Form der menschlichen Verbundenheit immer nur in einer kleinen Gruppe beginnen kann und zunächst nicht universell ist, so kann das Thema Universal Design vielleicht doch für die Idee der Solidarität sensibilisieren. Schließlich werden wir alle von der Hoffnung getragen, dass die gemeinsame Welt, in der wir leben, morgen ein Stück weit besser sein könnte, als sie es heute ist.

A comparison of the social state structures in Japan, Germany and the USA, in particular the different healthcare systems, makes it very clear that the question of solidarity remains, above all, a national concern. Despite human bonds over different cross-cultural groups, there remains this form of solidarity within a group of like-minded people or within a social state system. In this way, the problems of different countries might be similar. Unfortunately, the result is that they are already so similar to themselves that the question of solidarity has often been reduced to a pure problem of costs. It is necessary to solve this problem and moreover for the people to be sensitized to the idea of solidarity. Solidarity belongs to the heritage of a culture of humanity, the sympathy and the responsibility. Even if this form of human attachment can only begin in a small group and is not at first universal, perhaps the theme of universal design can actually sensitize the idea of solidarity. In the end, we are all carried by the hope that the world in which we live might be a bit better tomorrow than it is today.

Universal Design international – Perspektiven für die Produktentwicklung
Universal Design international – Perspectives for product development

Von Stephan Scheuer
By Stephan Scheuer

Zwei Entwicklungen haben in den beiden letzten Jahrzehnten dazu beigetragen, neue Produkt- und Systementwicklungen für einen breiteren Anwenderkreis besser zugänglich zu machen. In einigen Industrieländern haben Politik und Gesetzgeber über lange Zeit benachteiligte Gruppen wie Behinderte und chronisch Kranke als einen regulierungsbedürftigen Kreis identifiziert. Zunehmend sichtbar beziehen die Marktforschung und die Medien die demografische Entwicklung mehr zukunftsorientiert in ihre Analysen und Empfehlungen mit ein. Dabei haben sie auch die ältere Generation als eine Zielgruppe mit besonderen Bedürfnissen und Ansprüchen entdeckt.

Beide Beobachtungen lassen sich auf Fakten begründen. Nach Angaben der WHO sind derzeit etwa 650 Millionen Menschen von Behinderungen oder chronischen Krankheiten betroffen, das sind etwa 12 % oder jede achte Person der Weltbevölkerung. Auf Europa (EU) bezogen sind es anteilig ca. 50 Millionen Personen.

Nach Angaben der Vereinten Nationen liegt die Anzahl der „Über-60-Jährigen" derzeit bei etwa 600 Millionen Menschen weltweit und damit bei einem Anteil von etwa 11 %. Bereits jetzt liegen mit Japan, gefolgt von Deutschland, zwei große Nationen mit über 25 % Anteil an der Spitze und sind Vorläufer für die alternde Gesamtweltbevölkerung.[14]

In Summe sind es also 1,2 Milliarden Menschen, die nicht oder nur in Teilen die körperlichen, geistigen oder mentalen Fähigkeiten eines vollständig gesunden Menschen im mittleren Alter zur Verfügung stehen haben, mit einem deutlich anwachsenden Anteil.

Dies sollte Motivation genug sein, den Gedanken und die Prinzipien des „Universal Designs" als Grundrecht zu verankern und damit auch in den wichtigen Teilaspekten zu standardisieren. Die Umsetzung ist jetzt gefordert und bleibt die Herausforderung der nächsten Jahrzehnte.

Two developments have contributed in the last twenty years to improving the accessibility of new product and system developments for a broader circle of users. In some industrialized countries politics and lawmakers have long identified disadvantaged groups, such as the disabled and the chronically ill, as groups in need of extra regulation. Increasingly visible market research and media include the demographic development in their analyses and recommendations in a more future-oriented way. Thus, they have also discovered the older generation to be a target group with special needs and claims.

Both observations can be supported with facts. According to the WHO, there are nowadays about 650 million disabled or chronically ill people, that comes to approximately 12 % or every eighth person in the world's population. In the European Union there are approximately 50 million such people.

According to the United Nations, the number of the people over 60 years old is nowadays about 600 million worldwide, amounting to 11 %. Two large nations, Japan and Germany in a close second, are at the top of this list accounting for more than 25 % of the world's over 60 population, thereby leading the way for an aging world population.[14]

In all there are 1.2 billion people who partially or completely lack the physical, spiritual or mental abilities of a completely healthy middle-aged person, and this number continues to rise.

This should be motivation enough to anchor the ideas and principles of Universal Design as a fundamental right and to standardize important aspects. The implementation is demanded now and remains a challenge for the coming decades. A current topic for legislation is something the USA acted on many years ago with the "American with Disabilities Act" of 1990, a forerunner for the rights of disabled people, while in Asia, Japan followed with an initiative for disabled

Was die aktuelle Gesetzgebung betrifft, ist die USA mit der Anfang 1990 erlassenen Rechtsvorschrift der „Americans with Disabilities Act of 1990“ ein Vorreiter in der Breite, was die Rechtsansprüche benachteiligter Personengruppen betrifft. Während in Asien von Japan die Initiative für das Behindertengesetz von 1993 und das „Grundgesetz der Maßnahmen für die alternde Gesellschaft von 1995“ folgten, sind vergleichbare gesetzliche Rahmenvorgaben in Europa erst in den letzten Jahren begonnen worden. Das Behindertengleichstellungsgesetz in Deutschland aus dem Jahr 2002 gehört dazu. Eine wirksame europäische Gesetzgebung für Produkte existiert aber derzeit nicht.

Alle diese Initiativen zeigen langsam ihre Wirkung, wo es um neue öffentliche Infrastrukturen für die Bevölkerung geht, den Zugang zu Informationen oder die Beschäftigung in einem Arbeitsverhältnis. Im Vordergrund stehen drei Zielsetzungen, die in allen Lebensbereichen systematisch umzusetzen sind:

1. Benachteiligungen für eingeschränkte Benutzer beseitigen
2. Barrieren abbauen und Zugänge erleichtern
3. Selbstbestimmte Lebensführung ermöglichen

Am treffendsten aus internationaler Sicht greift die „Konvention über die Rechte behinderter Menschen“ der Vereinten Nationen aus 2007 das Thema auf.[15] Sie fordert den klaren Bezug zur Gestaltung von Produkten und bezieht damit erstmalig die herstellende und dienstleistende Industrie in einem globalen Umfeld ein. Nach Artikel 2 im Sinne des Übereinkommens, Absatz 5, heißt es:

„... bedeutet „universelles Design“ ein Design von Produkten, Umfeldern, Programmen und Dienstleistungen in der Weise, dass sie von allen Menschen möglichst weitgehend ohne eine Anpassung oder ein spezielles Design genutzt werden können. „Universelles Design“ schließt Hilfsmittel für bestimmte Gruppen von Menschen mit Behinderungen, soweit sie benötigt werden, nicht aus.“

Es folgt die klare Aufforderung an die verantwortlichen Kreise: „Forschung und Entwicklung für Güter, Dienstleistungen, Geräte und Einrichtungen in universellem Design, wie in Artikel 2 definiert, die den besonderen Bedürfnissen von Menschen mit Behinderungen mit möglichst geringem Anpassungs- und Kostenaufwand gerecht werden, zu betreiben oder zu fördern, ihre Verfügbarkeit und Nutzung zu fördern und sich bei der Entwicklung von Normen und Richtlinien für universelles Design einzusetzen.“

Die Konvention präzisiert auch den Zusammenhang zwischen „universellem Design“ und der „Zugänglichkeit“ in Artikel 9:

„Um Menschen mit Behinderungen eine unabhängige Lebensführung und die volle Teilhabe in allen Lebensbereichen zu ermöglichen, treffen die Vertragsstaaten geeignete Maßnahmen mit dem Ziel, für Menschen mit Behinderungen den gleichberechtigten Zugang zur physischen Umwelt, zu Transportmitteln, Information und Kommuni-

people in 1993 and the "Basic law on measures for the aging society" in 1995. In Europe, a comparable legal framework has only begun in the last few years. The German "Federal Act on the Equalization of Disabled Persons" of 2002 is part of it. Today there is no effective European legislation for products.

All these initiatives eventually prove their effectiveness when it comes to new public infrastructures, access to information or the activity level required of an employee. At the forefront are three objectives which should be systematically implemented in all areas of life:

1. Remove obstacles to disabled people
2. Minimize barriers and improve accessibility
3. Allow for independent living

The most visible international attention to this topic came from the United Nation's 2007 Convention on the Rights of Persons with Disabilities.[15] This demands a clear relationship between a product and its form and in so doing creates a unique global environment for manufacturers and the service industry. Specifically addressing sales, Article 2, paragraph 5 states:

"Universal Design' means a design of products, environments, programs and services in such a way that they can be used by all people to a great extent without adaptation or special design. 'Universal Design' does not exclude aid for certain groups of people with disabilities as far as they need them."

This clear request to the responsible parties follows: "Pursue or promote the research and development of goods, services, devices and equipment in Universal Design, as stated in Article 2, that meet the special needs of disabled people, requiring minimal to no adaptation or expense, to promote such products' accessibility and use and to keep them in line with the development of norms and guidelines for Universal Design."

The convention also specifies the connection between Universal Design and accessibility in Article 9:

"To enable persons with disabilities to live independently and participate fully in all aspects of life, States Parties shall take appropriate measures to ensure to persons with disabilities access, on an equal basis with others, to the physical environment, to transportation, to information and communications, including information and communications technologies and systems, and to other facilities and services open or provided to the public, both in urban and in rural areas."

With this the basic requirements and purposes of Universal Design and accessibility are generally and appropriately addressed. Today, both of these concepts find international application, but as far as laws and norms go, the concept of accessibility dominates internationally.

Die vollautomatische barrierefreie Toilettenanlage „WC Tronic 402“ der Firma Ströer wurde im Jahr 2008 erstmals in Dresden aufgebaut. Die Anlage wurde mit dem DIN-Siegel „Barrierefreies Planen und Bauen, barrierefreie Produkte – besser leben ohne Schranken“ zertifiziert.
The fully automatic barrier-free toilet system “WC Tronic 402” by the company Ströer was installed for the first time in Dresden in 2008. The system was granted the certified DIN seal “Barrier-free planning and building, barrier-free products – better living without barriers”.

kation, einschließlich Informations- und Kommunikationstechnologien und -systemen, sowie zu anderen Einrichtungen und Diensten, die der Öffentlichkeit in städtischen und ländlichen Gebieten offenstehen oder für sie bereitgestellt werden, zu gewährleisten.“

Damit sind die grundlegenden Anforderungen und die Ziele des „Universal Designs“ und der „Zugänglichkeit“ allgemein und global treffend adressiert. Heute finden beide Begriffe international Verwendung, wobei im gesetzlich und normativ geregelten Bereich der Begriff der Zugänglichkeit international dominiert. In Deutschland findet sich aber noch der Begriff der Barrierefreiheit nach § 4 Behindertengleichstellungsgesetz[16]:

„Barrierefrei sind bauliche und sonstige Anlagen, Verkehrsmittel, technische Gebrauchsgegenstände, Systeme der Informationsverarbeitung, akustische und visuelle Informationsquellen und Kommunikationseinrichtungen sowie andere gestaltete Lebensbereiche, wenn sie für behinderte Menschen in der allgemein üblichen Weise, ohne besondere Erschwernis und grundsätzlich ohne fremde Hilfe zugänglich und nutzbar sind.“

In Germany, though, there is still the concept of freedom of barriers as found in Section § 4 of the “Equalization of Disabled Persons Act”:[16]

“Buildings and other facilities, means of transport, technical articles of daily use, information processing systems, acoustic and visual sources of information and communication facilities as well as other specially designed areas of life are barrier-free if they can be accessed and used by people with disabilities in the ordinary way, without undue difficulty and as a general rule without help from others.”

The freedom from barriers is therefore a mark of quality and character for the accessibility of a product or system for different users and thereby is again seen as a decisive element of Universal Design. Even with these definitions and demands, a designer or product developer still has no concrete, handy product requirements which would allow for their most efficient implementation. Moreover, more knowledge is required, which is often hidden in other disciplines of product development. These are namely:

Die Barrierefreiheit ist damit eine Eigenschaft und Kenngröße für die Zugänglichkeit eines Produktes oder Systems für unterschiedliche Benutzer und damit wiederum als bestimmendes Element für ein universelles Design anzusehen.

Ein Designer oder Produktentwickler erhält aber mit diesen beschreibenden Definitionen und Forderungen noch keine konkreten greifbaren Produktanforderungen, die es ihm ermöglichen, diese Ziele effizient umzusetzen. Hierzu bedarf es weiterer Kenntnisse, die sich in anderen Disziplinen der Produktentwicklung verbergen; im Einzelnen sind dies:

- die Produktsicherheit,
- die Ergonomie und
- die Gebrauchstauglichkeit.

Daneben werden Faktoren des umweltbewussten und nachhaltigen Designs immer bedeutender.

Die Produktsicherheit ist die Basisanforderung an ein verkehrsfähiges Produkt und darf keinesfalls auf Kosten des universellen Designs leiden. Hierzu gab es Beispiele wie Leichtbauprodukte in Form von speziellen Steh- und Sitzhilfen, die einfach zugänglich und leicht zu bewegen waren, denen aber die Standfestigkeit fehlte und die damit ein hohes, nicht vertretbares Gefährdungspotenzial aufwiesen. Für ein effektives, effizientes und zukunftsweisendes Produktdesign sind die Aspekte der Ergonomie und Gebrauchstauglichkeit bereits heute und zukünftig noch stärker unumgänglich. In beiden zusammen liegt der Schlüssel für universelles Design. Sie sind eng miteinander verzahnt und bilden die antreibenden Werkzeuge des universellen Designs.

Produkte des universellen Designs sollten sich immer durch eine gute Ergonomie mit selbsterklärenden Produktmerkmalen und einer einfachen aufgabenangemessenen Bedienung auszeichnen. Denn nur so sind ein hohes Maß an Gebrauchstauglichkeit und damit effizienter Benutzung und das Ziel einer messbaren Benutzerakzeptanz überhaupt möglich. Vorrangig ist beim Universal Design aber die Zugänglichkeit für die effektive Nutzung.

Sowohl aus Hersteller- und Benutzersicht als auch aus Sicht einer unabhängigen Prüf- und Zertifizierungsorganisation, die neutral und eindeutig die Konformität mit den gesetzlichen Vorgaben, den Normen und Designprinzipien bestätigt, sind dafür robuste Messwerkzeuge und Bewertungsmethoden erforderlich.

Mit dem Verfahren der Kontextanalyse liefert die Gebrauchstauglichkeit oder Usability auch ein sehr geeignetes Instrument für die Erhebung konkreter Produktanforderungen. So etablieren sich gerade die Analyse der Kenntnisse über die vorgesehenen Benutzer und ihre Einschränkungen, die strukturierte Analyse der Aufgaben und der Umgebungen, in der das Produkt angewendet wird, als die zentralen Elemente des Anforderungsmanagements für eine Prüfung,

- product safety,
- ergonomics and
- usability of use.

In addition, considerations of the environmentally conscious and of lasting design become more and more significant.

Product safety is the basic requirement for a marketable product and may on no account suffer at the expense of Universal Design. There have been examples, and surely there will of lightweight products intended to aid sitting or standing and which could be moved easily but that lacked stability and thereby presented an inexcusably high danger. For effective, efficient and trendsetting product design, the aspects of ergonomics and usability are becoming more and more unavoidable. The key to Universal Design lies in both, which is why the disciplines safety, accessibility and usability constantly gain in importance. They are closely interlocked with each other and turn the wheels of Universal Design.

Products of Universal Design should always distinguish themselves with good ergonomics, self-explanatory product features and be easily able to accomplish its designed-for tasks. Only in this way can there be a high degree of usability, and therefore efficiency and a measurable user-acceptance. Accessibility for the actual use, however, is a priority of Universal Design.

Manufacturers, users and even independent testing and certification organizations who neutrally and unambiguously testify to the conformity of the statutory requirements, norms and design principles all see therefore that sturdy measuring tools and assessment methods are necessary.

Through context analysis, usability also delivers a very practical instrument to ascertain more concrete product requirements. Analysis of the knowledge about the intended users and their limitations and a structured analysis of the tasks and surroundings in which the product will be used, are just becoming central elements in testing, verifying and certifying Universal Design. In addition to the already known ergonomic requirements, the following should be taken into consideration:

1. Physical or physiological limitations of movement including walking, gripping and the senses of sight, hearing, taste and smell

2. Psychiatric and cognitive limitations to take in, process and interpret Information

The product range of application and surroundings should be taken into account in such a way that those products used in conjunction with public infrastructures meet all possible requirements and recommendations for use by all target groups. Unlike in the private sphere, no alternatives are available to the user in the public sphere.

Verifizierung und Zertifizierung des Universal Designs. Dafür sind über die bekannten ergonomischen Anforderungen hinaus zusätzlich zu berücksichtigen:

1. Physische oder physiologische Einschränkungen im Bereich der Bewegung und Motorik des Gehens, Greifens und der Wahrnehmungsfunktionen Sehen, Hören, Tasten und auch des Riechens

2. Psychische und kognitive Einschränkungen in der Informationsaufnahme, Informationsverarbeitung und Informationsinterpretation

Anwendungsbereiche und Umgebungen von Produkten sind in einer Weise zu berücksichtigen, bei der Produkte, die bei öffentlichen Infrastrukturen zum Einsatz kommen, möglichst alle Anforderungen und Empfehlungen für eine Benutzung durch alle Zielgruppen einhalten. Denn anders als im privaten Umfeld stehen hier meistens keine Alternativen für den Benutzer zur Verfügung.

Der Umsetzungstand dieser Forderungen in transparente technische Vorgaben ist global sehr unterschiedlich. Während in Ostasien mit Japan und auch partiell in Australien und in Nordamerika sowie in Nordeuropa bereits ein teilweise hoher praktischer Umsetzungsgrad erreicht wurde, ist dies in den anderen entwickelten Ländern nur eingeschränkt gegeben, darunter sind Südamerika, aber auch Mittel- und Südeuropa vertreten. In den Entwicklungsländern steht die Umsetzung noch ganz am Anfang.

Ein gutes Beispiel stellt die öffentliche Bereitstellung von Orientierungs- und Informationsdaten für Gehbehinderte und Sehbehinderte dar. So existieren in Nordamerika, Großbritannien, Japan und einigen anderen Staaten seit Mitte der 90er Jahre Gesetze und Normen, wie zum Beispiel die „ADA Standards for Accessible Design (28 CFR Part 36, Department of Justice)“[17], die in den USA oder auch in anderen Staaten wie Japan dafür gesorgt haben, dass öffentliche Wege und Haltestellen von Verkehrsmitteln wesentlich besser zugänglich geworden sind. Taktile und auditive Informationssysteme in Form von integrierten Signalen geben die erforderlichen Tast- und Akustikrückmeldungen für eine Orientierung und zur Navigationsunterstützung und sind Vorbild für andere Staaten. Im Zeitalter der Satellitennavigation sind aber auch fortgeschrittenere individuelle Produktlösungen denkbar, die der Anwender unabhängig vom Standort immer mit sich führt.

An vielen modernen öffentlichen Informationseinrichtungen scheinen die vorhandenen Kenntnisse und die internationalen, nicht immer ausgereiften Standards in fast allen Ländern noch weitgehend unberücksichtigt. Auch die Verordnung zur Schaffung barrierefreier Informationstechnik (BITV) in Deutschland ist an öffentlichen Plätzen, Haltestellen, Flughäfen und anderen öffentlichen Einrichtungen im Bezug auf effiziente Lösungen kaum sichtbar.[18]

The level of implementation of these demands in clear technical guidelines varies widely throughout the world. While in eastern Asia Japan, and also partially Australia, and in North America and northern Europe, there exists already a partially high level of implementation, it is only limited in the other developed countries, among them those of South America, as well as central and southern Europe. Among the developing countries, the level of implementation is just at the beginning.

A good example is the public supply of guidelines and data information for physically disabled people and the visually impaired. In North America, Great Britain, Japan and some other countries, laws and norms have existed since the mid-1990s, for example the "ADA Standards for Accessible Design (28 parts CFR 36, Department of Justice)"[17], which have provided in the USA, or also in other states like Japan, that public transportation become significantly more accessible. Tactile and audio information systems in the form of integrated signals give the necessary feedback for orientation and provide navigation support, and are models for other countries. In the age of satellite navigation, however, it is also conceivable that one day users will be able to direct themselves to and from anywhere.

In a lot of modern public information equipment, it seems the available knowledge and the not always mature international standards are still widely disregarded in nearly all countries. Also Germany's regulation for the creation of information technology free of barriers (BITV) is barely visible on public places, bus and tram stops, airports and other public areas requiring efficient solutions.[18]

The experiences of a testing and certification company such as TÜV Rheinland show that purchasers, operators and manufacturers of equipment must still develop their knowledge about the precise quantified state criteria for Universal Design.

Often there are deficits in the procurement of information and with it knowledge deficits which lead to products being used in surroundings and applications for which they were not intended. The application of a poorly reflective screen or display which is used outdoors and in daylight for information terminals is rather of little avail. The problem lies deeper, however, because most screens and displays are intended for the interior applications and are thus less expensive to purchase, something which is not helpful to reaching barrier free access to information.

Good Universal Design does not mark itself only by the observance of formal laws and general norms. It is the measurable state regulations and quantifiable norms, together with the above mentioned tools, which must be determined. Using the example of public information available at information terminals are the following examples:

Mit der elektrischen Bewegungsunterstützung „Servo-Drive" von Blum können Fronten ergonomisch vorteilhaft geöffnet werden.
With the electrical "Servo Drive" opening support system by Blum drawer and pull-outs can be opened ergonomically.

Auszüge mit „Servo-Drive", der elektrischen Öffnungsunterstützung von Blum, können bequem mit Hand, Knie, Hüfte oder Fuß geöffnet werden.
Pull-outs equipped with "Servo Drive", the electrical opening support system by Blum, can be opened conveniently by hand, knee, hip or foot.

Die Erfahrungen in einem Prüf- und Zertifizierungsunternehmen wie dem TÜV Rheinland zeigen, dass sowohl Beschaffer und Betreiber als auch Hersteller der Einrichtungen ihre Kenntnisse über die genauen quantifizierten Beschaffenheitskriterien für ein universelles Design erst noch weiterentwickeln müssen.

Oft sind es Defizite in der Informationsbeschaffung und damit Wissensdefizite, die dazu führen, dass Produkte für Umgebungen und Anwendungen eingesetzt werden, für die sie nicht konzipiert sind. So ist der Einsatz eines schlecht entspiegelten Bildschirms oder Displays, das im Außenbereich und bei Tageslicht in Informationsterminals eingesetzt wird, weniger hilfreich. Das Problem liegt aber tiefer, da die meisten Bildschirme und Displays für Innenanwendungen bei niedriger Beleuchtungsstärke konzipiert und damit günstiger in der Anschaffung sind, was wenig zielführend für einen barrierefreien Informationszugang ist.

Gutes universelles Design zeichnet sich demnach nicht alleine durch Einhaltung von formalen Gesetzen und allgemeinen Normen aus. Es sind die messbaren Beschaffenheitseigenschaften und quantifizierbare Normgrößen, die mit den oben bereits genannten geeigneten Werkzeugen zu ermitteln sind. Am Modell der öffentlichen Informationsbereitstellung auf Informationsterminals sind dies beispielhaft:

1. Legibility of the display, specifically addressing:
– representation of information in characters and symbols larger than 8 mm with 1 m of visual distance or at least 12 mm with a maximum of 2 m of visual distance

– effective reflexion of the display in light, so that contrasts of at least 5:1 to 10:1 are guaranteed

– sufficient brightness of the displays for outdoor applications, e.g., as a rule more than 400 cd/m² light density with emissive display technologies

– wide visual angle in the display, for example, greater than 140 degrees the above mentioned contrast relations

2. Use of the operating equipment
– accessible arrangement of all operating devices so that they can be reached by users of different heights as well as by seated users, that is at least with consideration of the normal reaching distances in standing and seated positions

– the operating elements must be sized appropriately and be arranged with sufficient distance in order to be served by persons with

1. Lesbarkeit der Anzeige im Bezug auf die Eigenschaften:
– Informationsdarstellung von Zeichen und Symbolen größer als 8 mm im Bezug auf 1 m Sehabstand oder mindestens größer als 12 mm bei maximal 2 m Sehabstand

– wirksame Entspiegelung der Anzeige bei Lichteinfall, damit Kontraste von mindestens 5:1 bis 10:1 gewährleistet sind

– ausreichende Helligkeit der Displays für Außenbereichanwendungen, z. B. mit in der Regel mehr als 400 cd/m² Leuchtdichte bei emissiven Anzeigetechniken

– geringe Blickwinkelabhängigkeit der Anzeige, z. B. größer als 140 Grad für die oben genannten Kontrastverhältnisse

2. Benutzung der Bedieneinrichtungen
– Anordnung aller Bedienteile zur Bedienung in der Art, dass diese von Benutzern unterschiedlicher Körpergröße stehend und von Menschen, die sitzen, erreichbar sind, d. h. mindestens Berücksichtigung der normativen Greifbereiche im Sitzen und Stehen.

– Die Bedienelemente müssen angemessen groß und mit ausreichendem Abstand angeordnet sein, um von Personen mit eingeschränkter Motorik bedient werden zu können, d. h., normativ vorgesehene Abstände zwischen Tasten oder Schaltern sind in der Regel deutlich um ca. 20-30 % zu vergrößern.

– Eine ausreichende Dimensionierung der Bedien- und Stellteile gemäß den ergonomischen Grunddaten für Hand- und Fußbetätigung ist zu berücksichtigen.

– Leichte, einfache Betätigung der Bedienelemente gemäß den anthropometrischen Grunddaten für Kräfte und Drehmomente, wobei zu berücksichtigen ist, dass die normativen Werte zwischen 30 % und 50 % niedriger anzusetzen sind, z. B. für die Auslegung und Betätigung einer Notruftaste.

– Die Möglichkeit zur akustischen Ausgabe der Informationsanzeige ist zu berücksichtigen mit möglichst anpassbarer Lautstärke.

– Die Möglichkeit zur akustischen Steuerung und Navigation der Bedienfunktionen ist zu berücksichtigen, z. B. Mikrofon- und Spracheingabeoption.

– Die Gestaltung interaktiver Dialoge ist gemäß den Dialogprinzipien für Softwareanwendungen durchzuführen, wie sie in den Normen ISO 9241-110 und ISO 9241-171 beschrieben sind:

– Der Benutzer muss genügend Zeit haben, um seine Aufgaben zu erledigen.

limited motor capacity, i.e. normative intended distances between keys or counters are, as a rule, to increase approximately 20-30 %

– sufficient dimensions of the operating devices and their parts must be taken into consideration as per the basic ergonomic data for hand and foot operation

– light, easy operation of the controls according to the basic anthropometric data for forces and torques and is to be taken into consideration so that the normative values are between 30 % and 50 % lower, e.g., for the layout and operation of an emergency call key

– the possibility that the acoustic element of the information display can have an adjustable volume setting

– the possibility to adjust the acoustic controls and navigation of the operating functions is to be taken into consideration, e.g., microphone and language input options

– the creation of interactive dialogs is to be carried out according to the dialog principles for software applications as they are described in the norms ISO 9241-110 and ISO 9241-171:

– the user must have enough time to perform the tasks

– the user must have the possibility to quit the whole process at any time and to be able to reject the given data

3. Location choice
– as regards mounting place, mounting height, inclination, illumination and climatic influences the choice of location has to be made in such a way that users may have access to the dialogues with or without any additional means

– the locations may not be in remote places which require an additional expenditure of energy and added stress for the user

The practical example of public points of information clearly shows where the need for action lies for visible solutions of Universal Design. It is not chosen without a reason, because most of the requirements named here can be transferred to the private sector.

If we look at communication technology in the household, there are comparable requirements. These include a phone with display and operating keys, a microwave with display and program settings for heating and cooking times, or the type of washing machine interface to laundry and the choice of washing programs on a labelled or electronic display.

In addition, door or window handles with an average 20-50 % reduced operating strength compared with the normative values, as well as slip-proof and cleaning-friendly materials and a 10-30 %

– Der Benutzer muss die Möglichkeit haben, den gesamten Vorgang jederzeit abbrechen und die eingegebenen Daten verwerfen zu können.

3. Standortwahl

– Die Auswahl des Standortes ist im Bezug auf Anbringort, Anbringhöhe, Neigung, Beleuchtung und Klimaeinflüsse so auszulegen, dass von Benutzern mit und ohne Hilfsmittel ein einfacher Zugang zu den Dialogen möglich ist.

– Die Standorte dürfen nicht an abgelegenen Stellen liegen, die einen zusätzlichen Energieaufwand und eine zusätzlich messbare Belastung für den Benutzer darstellen.

Das Praxisbeispiel der öffentlichen Informationssysteme zeigt transparent, wo der Handlungsbedarf für sichtbare Lösungen des Universal Designs liegen. Es ist nicht ohne Grund gewählt, denn die meisten der hier genannten Anforderungen lassen sich auf den privaten Bereich übertragen.

Betrachten wir die Kommunikationstechnik im Haushalt, gelten vergleichbare Anforderungen. Diese gelten für ein Telefon mit Display und Bedientasten, eine Mikrowelle mit Anzeige und Stellreglern für die Wärmeleistung und Garzeit oder das Interface einer Waschmaschine zur Bestückung mit Wäsche und der Auswahl der Waschprogramme auf einer beschrifteten oder elektronischen Anzeige.

Auch der Tür- oder Fenstergriff mit im Durchschnitt 20-50 % reduzierter Betätigungskraft gegenüber den normativen Werten sowie rutschfeste und reinigungsfreundliche Materialieneigenschaften und eine um 10-30 % niedrigere Anordnung in der Höhe sind eindeutig messbare Eigenschaften. Alle diese Beispiele stellen für den nicht mit Einschränkungen behafteten Benutzer in der Regel ein Komfortmerkmal dar, für den mit physischen und physiologischen Problemen ausgestatteten Benutzer aber sind sie erforderlich und im Notfall lebensrettend.

Die Entwicklung barrierefreier und auf der nächsthöheren Ebene universell nutzbarer Produkte und Systeme benötigt hierzu eine international intensive, vernetzte und effiziente Standardisierung mit quantifizierten Produktanforderungen.

Die Vereinten Nationen und die Gesetzgeber der international führenden Nationen haben aus heutiger Sicht einen ausreichenden Rahmen vorgegeben. Alle Gesellschaftsteile und insbesondere die Anwender müssen dies in der Praxis mit mehr Präzision einfordern. Dann wird die Industrie für die Herstellung und Bereitstellung infrastruktureller Systeme und auch für Konsumgüter verstärkt mit einem universellen Design zum Nutzen für alle Beteiligten in dynamischer Weise folgen.

reduced height are clearly measurable qualities. As a rule, all of these examples represent levels of comfort for the non-disabled user, and for those with physical and physiological problems they are necessary and sometimes even life-saving.

The development of barrier free products and ever improved levels of universally usable products and systems requires an intense, coordinated and efficient international standardization with quantified product requirements.

The United Nations and the legislators of the leading international nations have provided a sufficient, modern framework. All members of society, especially the users, must call for this to be put into practice with more precision. Then the industry producing and supplying infrastructure systems and consumer goods will follow with a heightened Universal Design which will bring benefit for all involved parties in a dynamic way.

Vom Universal Design zum Universal Home
From Universal Design to Universal Home

Von Eckhard Feddersen und Insa Lüdtke
By Eckhard Feddersen and Insa Lüdtke

Multifunktional: Die Fensterbank lädt ein zum Abstellen, Rasten und Plauschen.
Multifunctional: The windowsill invites for stalling, resting and chatting.

Orientierung: Der Etagenflur gliedert sich in „private" und „öffentliche" Raumzonen. Farbige Paneele und das Fußbodenlayout betonen die Wohnungseingänge.
Orientation: The floor is divided in "public" and "private" zones. Coloured panels and the layout of the flooring highlight the entries of the flats.

Wohnen ist ein lebenslanger „Prozess". Damit werden künftig nicht nur die Parameter für die Produktion von alltäglichen Gebrauchsgegenständen, sondern auch für die Konzeption von Häusern und Städten den Ansprüchen sowohl von Kindern als auch alten Menschen entsprechen müssen. Entstanden zuerst in den USA, propagiert „Universal Design" eine Formgebung von Alltagsgegenständen und der Umweltgestaltung, die für jedermann nutzbar ist. Diese Herangehensweise etabliert einen hohen Qualitätsanspruch an Materialien, Design und Komfortstandard und integriert Gegenstände mit Hilfe-Impetus, wie etwa Haltegriffe, in die Umgebung.

Den Begriff „Universal Design" kann man zudem im Sinne von „universell nutzbar" definieren: Die flexible und vielseitige Nutzung des Wohnumfeldes (z. B. terrassierte Platzgestaltung zum Sitzen für Ältere und zum Spielen für Kinder) und von Alltagsgegenständen (z. B. Handtuchhalter im Bad, der gleichzeitig Haltegriff für ältere Menschen ist) ist eines der wesentlichen Grundprinzipien des Universal Designs. Bewusst wird bei der Gesamtkonzeption eines Wohnquartiers auf besondere Anpassungen für „spezielle" Gruppen (z. B. Menschen mit Behinderungen oder Kinder) verzichtet, um eine Ausgrenzung und Stigmatisierung zu verhindern und vielfältige Wahlmöglichkeiten und Individualitäten zuzulassen.

Auch wenn im hohen Alter (z. B. Demenz) und im Kindesalter die intellektuellen Fähigkeiten nicht mehr bzw. noch nicht ausgeprägt sind, sollten im Wohnumfeld für alle Bewohner die Grundbedürfnisse von Sicherheit und Orientierung wie auch Intimität und Gemeinschaft sinnlich erfahrbar sein, z. B. über einen einprägsamen und stets wiederkehrenden Materialkanon und den Einsatz von differenzierten Lichtstimmungen. Dies sind alles Beispiele für das Prinzip „easy going". Universal Design meint aber auch die Disposition, also eine geschickte Anordnung der Sanitärobjekte, Raumdispositionierung und -proportionierung. Etwa die Unterfahrbarkeit des Waschtisches erfordert eine Großzügigkeit, die sich in der Schweiz bereits durchgesetzt hat. Mieter setzen diese Qualität bei jeder Wohnungssuche bereits von vornherein voraus.

Die Zugänglichkeit ins Freie ist eine andere Form der Großzügigkeit. Knapp ein Drittel von 1.600 Befragten in acht deutschen Städten wählten nach einer Umfrage des Potsdamer Instituts für Soziale Stadtentwicklung das (römische) „Gartenhofhaus" zur beliebtesten Wohnform. Bei dem ebenerdigen Haustyp, bei dem die Wohnräume einen Garten oder Gartenhof umschließen, sei die Verbindung von naturbezogenem Öffnen der Wohnung und familiärer Abgeschlossenheit entscheidend. Diese Erkenntnisse auf den Geschosswohnungsbau übertragen machen deutlich, wie wichtig ein hochwertiges – grünes – Wohnumfeld bzw. das Bedürfnis nach einem Balkon für Bewohner sein kann. Auch ein – barrierefreier – Austritt auf den Balkon kann schon solch ein sinnliches Erlebnis sein, wenn man etwa die Temperaturdifferenz spüren kann. „Wohnen" sollte zukünftig noch viel mehr die komplementäre Verbindung von Innen- und Außenraum beinhalten.

Living is a lifelong "process". In the future, not only the parameters for the production of articles of daily use, but also for the conception of houses and cities will thus have to meet the demands of both children and the elderly. "Universal Design" – at first developed in the USA – propagates a design of articles of daily use and the living environment which is utilisable for everyone. This approach establishes a high-quality demand in terms of materials and design as well as standards of comfort, and integrates objects with help impetus such as, for example, grab handles into the environment.

The term "Universal Design" can also be defined in the sense of "universally utilisable": the flexible and versatile use of the residential environment (e.g., terraced setting of spaces with seats for the elderly and playgrounds for children) and other articles of daily use (e.g., towel rails in the bathroom which serve as hand rails for the elderly at the same time) is one of the essential and fundamental principles of Universal Design. In the overall conception of residential quarters, a deliberate effort is made to do without certain adjustments for "special" groups (e.g., handicapped people or children) so as to avoid exclusion and stigmatisation, and to allow for multifarious choice options and individualities.

In the residential environment, basic needs in terms of security and orientation as well as both intimacy and community should be sensually experiencable for all residents, even if intellectual capabilities are no longer so pronounced due to old age or dementia, or not yet developed as in the case of children. This can be implemented, for example, with a memorable and steadily recurring material canon and the use of differentiated mood lighting. These are all examples for the so-called "easy going" principle. However, Universal Design also refers to the disposition, e.g., an adept arrangement of sanitary objects as well as space disposition and proportionation. As an example, the headroom below a wash basin demands a certain generosity in terms of space – something that has already gained wider acceptance in Switzerland. Tenants looking for residences are already including this particular quality in their search criteria.

Access to the outdoors is another form of generosity. In a survey conducted by the Potsdamer Institut für Soziale Stadtentwicklung (Potsdam Institute for Social Urban Development), almost a third of 1,600 interviewees in eight German cities voted for the (Roman) "Garden Courtyard House" as the most popular form of living environment. This particular house type with ground level access and residence quarters enclosing a garden or garden courtyard, represents the crucial combination of natural open residential spaces and familial seclusion. Transferring these findings to multi-storey apartment construction elucidates the importance of a high-quality – green – living environment and/or the need for a balcony for residents. A – barrier-free – exit out to the balcony can already be such a sensual experience, for instance, if one can feel the difference in temperature. In the future, "living" should include, much more than heretofore, the complementary connection of in- and outdoors.

WOHNLÖSUNGEN SOLUTIONS FOR LIVING

In einer alternden Gesellschaft möchten die meisten Menschen in ihrer „gewohnten" Umgebung alt werden. Das bedeutet in Zukunft eine Stärkung der Wohnquartiere mit hochwertigen Angeboten von Räumlichkeiten für Gemeinschaft und Service. Dafür werden zwischen der freien Wohnungswirtschaft Kooperationen mit Trägern in der Altenhilfe entstehen müssen.

Ein eingebautes Möbel, das sowohl Sitzbank, Regal, Briefkasten oder Ablage sein kann, lädt im luftigen und hellen Atrium-Wohnflur zum Verweilen und Kommunizieren ein. Die überschaubare Anzahl an Nachbarn gibt dem Bewohner ein Gefühl von Sicherheit und Vertrautheit. In den Wohnungen vermittelt eine individuell gestaltbare „Wohnwand" (Kamin, Regal, Schrank) das Gefühl von Geborgenheit. Gleichzeitig öffnen sich die Räume mit großen Fenstern zum Balkon und Garten. Hochwertige Materialien vermitteln dem Bewohner Sinnlichkeit, Authentizität und Selbstwert.

Universal Design im Wohnungsbau meint eben nicht standardisierte Lösungen bzw. solche „für alle Fälle", sondern eine Abwägung dessen, was für den Einzelnen im Detail das Beste ist. In der Schweiz wird seit geraumer Zeit der „anpassbare Wohnungsbau" postuliert. Unsere Nachbarn fassen das Leben als Kontinuum auf, bei dem alle Lebensphasen fließend ineinander übergehen: Man sieht sich – einmal als „Senior" – nicht plötzlich aus dem Kreis der „Menschen" ausgeschlossen und gezwungen, nach einer geeigneten Wohnumgebung Ausschau zu halten. Diese Entstigmatisierung bedeutet, keine besonderen Auswahlkriterien zu Grunde zu legen, sondern vielfältige Wahlmöglichkeiten und Individualitäten zuzulassen: Nicht Unterordnung diktieren, sondern Prozesse evozieren. Den Pauschaltypus „älterer Mensch" oder „Rollstuhlbenutzer" gibt es nicht. Regularien und Planungsvorgaben sollten Spielraum für Varianten zulassen, differenzierte Nutzerprofile ermöglichen und vor allem Ausgrenzungen vermeiden. Stattdessen unterstützt die Architektur unterschiedliche Lebensentwürfe – unabhängig vom Alter der Bewohner.

Die Markenindustrie hat die „Best Ager" bereits als neue und solvente Zielgruppe erkannt. Nach Berechnungen des Deutschen Instituts für Wirtschaftsforschung verfügen die über 60-Jährigen bereits heute über eine Kaufkraft von rund sieben Milliarden Euro. Ihr Vermögen ist damit viermal höher als das der 20-Jährigen. Diverse Trendstudien geben Produktentwicklern Auskunft über die Befindlichkeiten und Bedürfnisse der „Silver Consumer", die 2020 fast ein Drittel der Gesamtbevölkerung ausmachen werden.

Mit diesen fundierten Kenntnissen ging in den letzten Jahren die sukzessive Markterschließung durch Alltagsprodukte – von der Telefonie über die Automobilindustrie bis hin zur Sanitärbranche – einher: Branchenübergreifend liefern alle Produkte mit ihrer Gestaltung Komfort und Sicherheit als Feature bzw. Mehrwert selbstverständlich mit. Der integrierende Ansatz einer „barrierefreien Umwelt für alle", der letztlich einmal vom Bauen her kam, wurde von der Herstellerindustrie überholt. Vermutlich liegt das an der dem Bauprozess in-

In an aging society, most people want to grow old in their "habitual" living environment. In the future, this means placing emphasis on residential quarters with high-quality offers of premises for community building and service. For that purpose, cooperation between the residential trade and industry on the one side, and legal entities assisting the elderly on the other, will have to come into being.

A built-in piece of furniture which can serve as seating bench, shelf, letterbox, or rack will invite residents to linger around in the airy and bright atrium and engage in communication. The small and select number of neighbours provides a feeling of security and familiarity to the residents. An individually configurable "wall unit" (fireplace, bookshelf, closet) in the residences will convey a feeling of cosiness and protection. At the same time, the rooms feature large windows out to the balcony and the garden. High-grade materials convey an atmosphere of sensuality, authenticity, and self-worth.

Universal Design in residential construction does not mean standardised and/or "all-purpose" solutions but a fine balance of that which, in detail, is the best for the individual person. In Switzerland, "adaptable residential construction" has been postulated for a considerable time already. Our neighbours understand life as a continuum in which all phases of life fluently blend into each other: one does not perceive oneself – as a "senior" – as excluded from the circle of "people", forced to look out for suitable residential quarters. This destigmatisation means not to take specific selection criteria as a basis, but rather allow for diverse and multiple choice options and individualities: it is not about dictating subordination but about evoking certain processes. There doesn't exist anything like the stereotypes "elderly" or "wheelchair user". Rules and planning specifications should give leeway for alternatives, allow for differentiated user profiles, and, most of all, avoid social marginalisation. Instead, "Universal Design" architecture supports different conceptions of life – independent from the residents' age.

The brand industry has already recognised the "best agers" as a new and solvent target group. According to calculations of the Deutsches Institut für Wirtschaftsforschung (German Institute for Economic Research), the over-60-year-olds have a combined purchasing power of about 7 billion euros. Their fortune is thus four times as large as that of the 20-year-olds. Various trend studies provide information to product developers about the needs and sensitivities of the "silver consumers" who will amount to almost one third of the total population by 2020.

In recent years, the successive tapping of new markets for articles of daily use – from telephony and the automotive industry all the way to the sanitary sector – was accompanied by substantiated knowledge gained in these fields: across all sectors, all products provide, as a matter of course, comfort and safety as a feature and/or added value with their design. The integrating approach of a "barrier-free environment for everyone", originally conceived by the construction industry, has been adopted and refurbished by the manufacturing

newohnenden Vielzahl an Akteuren und der damit verbundenen hohen Komplexität der Entscheidungsfindung.

Vermutlich wird sich in Zukunft der Druck in Hinblick auf Marktfähigkeit von Immobilien von Seiten der Bauherren bzw. der Projektentwickler und deren Marketingexperten verstärken und so den Ansatz des Universal Designs im Sinne des Universal Home weiter in die Architektur bringen.

Die heute über 50-Jährigen befinden sich im Übergang von einer defensiven hin zu einer selbstbestimmten Lebensweise. Wie auch der Rest der Gesellschaft zerfällt diese Zielgruppe immer mehr in Mikrosegmente. Noch nie waren die Lebensstile derart komplex und widersprüchlich. Außerdem vergrößert sich die Zeitspanne des „aktiven Alters". Erst ab dem 80. Lebensjahr tritt im Allgemeinen die vierte Lebensphase ein. Dann steigt das Risiko für demenzielle Erkrankungen und Pflegebedürftigkeit stark an. Trotz Gebrechlichkeit möchten die meisten Menschen im Alter jedoch heute in ihrer vertrauten Umgebung wohnen bleiben.

Wohnangebote müssen sich künftig immer mehr als bunte Angebotspalette ausdifferenzieren, um den heterogenen Bedarf zu befriedigen. Gerade die älteren Bewohner würden künftig vermehrt wohnungsnahe Dienstleistungen nachfragen, um einen Umzug ins Heim zu vermeiden. Dabei werden die Akteure der Wohnungs- und Pflegewirtschaft Hand in Hand gehen müssen, um integrierte Quartierskonzepte im Universal Design zu schaffen: vom Kindergarten bis zur Demenz-WG. Bewohner wie Betreiber könnten so beiderseits von Synergieeffekten profitieren: Seniorentreff und Kindergarten beleben sich gegenseitig, Bewohner einer Service-Wohnanlage könnten die Annehmlichkeiten des benachbarten Hotels nutzen.

industry. Presumably, this is because of the large number of participants inherently involved in the construction process, and the high degree of complexity of the related decision-making process.

In the future, the pressure exerted by building owners and/or project developers and their marketing experts with regard to the marketability of real estate objects will perhaps increase and thus further introduce the Universal Design approach in the sense of the Universal Home into the field of architecture.

Those who are over 50 today find themselves in a transition period from a defensive to a more self-determined way of life. Just like the remainder of society, this target group disintegrates into microsegments more and more. Never before have the different kinds of lifestyles been so complex and contradictory. In addition, the "active age" time span is considerably prolonged. In general, the so-called "fourth phase of life" does not start before the age of 80. From then on, the risk of old-age diseases such as dementia, and the general nursing care dependency is strongly increased. However, despite infirmity, most of the elderly prefer to stay in their habitual residential environment.

In the future, accommodation services have to allow for a broad range of choices so as to meet the heterogeneous demand. The elderly in particular will increasingly request services in their immediate residential environment in order to avoid moving into a retirement home. For that purpose, the responsible parties in the housing industry on the one side, and the nursing care industry on the other, will have to closely cooperate so as to create integrated residential concepts in the sense of Universal Design – from kindergarten all the way to the dementia residential community. Residents as well as operators could thus both benefit from synergy effects: social gatherings of the elderly here and the kindergarten there can mutually stimulate themselves, residents of a service-oriented housing complex could benefit from the conveniences of the neighbouring hotel.

Natürlich altern ohne alt zu sein – Technische Lösungen für soziale Herausforderungen?

Aging naturally without being old – Technical solutions for social challenges?

Von Reiner Wichert, Thomas Norgall
By Reiner Wichert, Thomas Norgall

1. Ausgangssituation und Motivation

„Alt werden, aber nicht alt sein" – der demografische Wandel führt dazu, dass dieses Motto mehr und mehr zum Leitthema unserer Gesellschaft wird.[19] Gesundheit und Bevölkerungsentwicklung werden in der gesellschaftlichen Diskussion der kommenden Jahre dominierende Themen sein. Rückläufige berufstätige Bevölkerungsanteile und Überalterung stellen insbesondere die Renten- und Pflegeversicherungssysteme vor bisher ungekannte Probleme. Dabei spielt die Aufrechterhaltung angemessener gesundheitlicher Versorgung der älter werdenden Bevölkerung eine übergeordnete Rolle, insbesondere angesichts stetig steigender Kosten im Gesundheitswesen sowie des absehbaren Fehlens von erforderlichem Personal in den kommenden Jahren. Direkt Betroffene, der soziale Dienstleistungssektor im Allgemeinen und das Gesundheitswesen im Besonderen sind damit zunehmendem Druck ausgesetzt, neue Versorgungskonzepte für die alternde Gesellschaft zu entwickeln.

Im Gesundheitswesen haben unter Maximen wie „ambulant vor stationär", „Prävention vor Kuration", „Rehabilitation vor Pflege" entsprechende Veränderungen bereits begonnen. Absehbar ist dabei, dass sich (i) die gewünschten und erforderlichen Unterstützungsleistungen in privater häuslicher Umgebung nur mit Technikeinsatz bezahlbar und praktikabel gestalten lassen und (ii) parallel zum (und teilweise hervorgerufen durch) häuslichen Einsatz von Technik weiterer Bedarf an professioneller Unterstützung – z. B. Überwachungsleistungen in telemedizinischen Zentren – entsteht, dessen Umfang und Kosten wiederum nur durch entsprechenden Technikeinsatz begrenzt werden können.[20]

Unter dem Begriff „Ambient Assisted Living" (AAL) werden derzeit Konzepte, Produkte und Dienstleistungen erforscht und entwickelt, die es mittels Informations- und Kommunikationstechnologie Menschen mit besonderen Bedürfnissen ermöglichen, länger selbstbe-

1. Starting point and motivation

"Growing old, but not being old" – the demographic shift leads to the fact that this motto more and more becomes a key topic in our society.[19] Health and population development will be dominant topics in the social discussion of the coming years. Due to a declining working population and an increase in the percentage of old people, pension schemes and health care insurance systems in particular are faced with previously unknown problems. In this context the maintenance of adequate health care for the aging society plays an overriding role, particularly in the face of constantly increasing health care costs as well as the foreseeable shortage of required staff in the coming years. People affected directly, the social services sector in general and health care in particular are thus under increasing pressure to develop new care concepts for the aging society.

In the health care sector corresponding changes have already begun under maxims such as "outpatient treatment before inpatient treatment", "prevention before cure" and "rehabilitation before care". It is foreseeable that (i) the desired and required support services in a private home environment can only become affordable and practical with the use of technology and (ii) parallel to (and partly caused by) the use of technology in the home, an additional need for professional support – such as monitoring services in telemedical centres – arises, the extent and cost of which can in turn only be limited by the use of corresponding technology.[20]

The term "Ambient Assisted Living" refers to the current research and development of concepts, products and services, which with the help of information and communication technology enable people with special needs to independently live in their familiar environment for longer. This corresponds to the prevailing wish of elderly people to remain in their familiar living environment as long as possible even in the face of increasing need for help and care.[21]

stimmt in ihrer gewohnten Umgebung zu leben. Dies entspricht dem überwiegenden Wunsch älterer Menschen, auch bei steigendem Hilfe- und Pflegebedarf so lange wie möglich in ihrer gewohnten Lebensumgebung zu bleiben.[21] Offenkundig wird dadurch die Lebensqualität der Betroffenen signifikant gesteigert bzw. erhalten.[22] Um dies zu erreichen, beziehen heutige technische Lösungen fall- bzw. personenbezogene Informations-, Überwachungs- und Entscheidungsunterstützungssysteme ein, die auf der Integration gesundheitsbezogener Daten und Dienste für die betreffenden Personen basieren.

2. Beispiele für Ambient Assisted Living

Jüngste Entwicklungen haben gezeigt, dass Forschung und Industrie durchaus in der Lage sind, Produkte für Ambient Assisted Living zu entwickeln, die bereits in wenigen Jahren vermarktet werden können. Insbesondere haben Forschungsvorhaben in den Bereichen Wohnen, häusliche Pflege, Rehabilitation, Integration von pflegebedürftigen oder behinderten Personen zu einschlägigen Ergebnissen geführt.

Vor dem Hintergrund der geschilderten Ressourcen- und Kostenproblematik im Gesundheitswesen charakterisiert der Begriff „Personal Health" einen Paradigmenwandel, welcher zu personenzentrierten, individualisierten Formen der Prävention, Diagnostik, Therapie und Pflege führt. Dieser Wandel wird maßgeblich von Entwicklungen in Mikroelektronik, Sensorik, Funktechnik, Bio- und Nanotechnologie ermöglicht und motiviert. Er konkretisiert sich im intensiven Einsatz von Informationstechnologie und Telematik zur personenbezogenen Integration digitaler Patientendaten, meist in Verbindung mit verschiedenen Formen von Telemedizin, insbesondere aber in innovativen personalisierten Telemonitoring-Systemen.

Um per Telemonitoring etwa Mehrkanal-EKGs bei hohem Tragekomfort und mit möglichst geringer Belastung für den Patienten zu erfassen, können Sensormodule mit hautverträglichen Trockenelektroden in körperangepasste Sensorshirts integriert werden. Die verdeckte Führung der Elektrodenleitungen und die Kapselung der Elektronik verhindern dabei nicht nur unerwünschten Kontakt mit der Haut des Patienten, vielmehr kann die Elektronik – mit Ausnahme des über einfach lösbare Steckverbindungen angeschlossenen Akkus – mit dem Shirt mitgewaschen werden. Die automatische Signalauswertung und Ereigniserkennung kann dabei weitgehend innerhalb des körpernahen Systems erfolgen. Die äußerst wichtige Blutsauerstoffsättigung wird – über die Pulswellenmessung der reflektierten oder durchdringenden Anteile des in Blutgefäße eingestrahlten Lichts definierter Wellenlängen – bis heute überwiegend an Finger oder Ohrläppchen erfasst. Dies wird jedoch als sehr störend empfunden und verhindert Langzeit- oder kontinuierliche Überwachung im Alltag. Zukünftige Lösungen, die sich etwa bequem als Armbanduhr oder Armreif am Handgelenk tragen lassen, werden hier neue mobile Einsatzfelder eröffnen. Die erfassten Sensordaten

This either significantly increases or maintains the quality of life of those affected.[22] In order to achieve this, today's technical solutions incorporate case-related or personal information systems as well as monitoring and decision support systems, which are based on the integration of health data or services for the respective person.

2. Examples of Ambient Assisted Living

Recent developments have shown that research and industry are indeed capable of developing products for Ambient Assisted Living, which can be marketed in a few years. In particular in the fields of living, domiciliary care, rehabilitation and the integration of people requiring care or with disabilities, research projects have led to relevant results.

Against the background of the resources and cost problems described above the term 'Personal Health' characterises a paradigm shift, which leads to person-centred and individual forms of prevention, diagnostics, therapy and care. This shift is substantially enabled and motivated by developments in micro electronics as well as sensor, radio, bio- and nanotechnology. It becomes apparent in the intensive use of information technology and telematics for the integration of digital, personal patient data, usually in combination with different forms of telemedicine, but in particular in innovative, personalised telemonitoring systems.

In order to record multi channel ECGs via telemonitoring while at the same time providing high wearing comfort with the smallest amount of stress for the patient, sensor modules with skin-friendly dry electrodes were integrated in sensor shirts adjusted to the body. The concealed electrode cables and the enclosed electronics not only prevent contact with the patient's skin, but the electronics can – with the exclusion of the battery, which is connected via an easy-to-use plug connection – simply be washed together with the shirt. The automatic signal interpretation and event detection can mostly take place within the system that is located close to the body. To date, the extremely important blood oxygen saturation has been mainly measured – via the pulse wave measurement of the reflected or permeating parts of light of defined wave lengths shone into blood vessels – at the finger or the earlobe. This, however, has been very annoying for wearers and prevents long-term or continuous daily monitoring. Future solutions, which will be comfortable to wear on the wrist in the form of wristwatches or bracelets, will open completely new areas of mobile application. The data captured via the sensors are usually transferred to a central unit located close to the patient's body or in their immediate environment to enable further processing, transmission or storage. There, for instance, the pulse wave transmission time can be calculated based on the combination of ECG and pulse wave data. Measured and calculated data and information are forwarded via a communication network to a server, which analyses them and in case of an emergency gets in contact with a service provider or doctor depending on the level of urgency.

werden üblicherweise zu Weiterverarbeitung und Weiterleitung bzw. Speicherung an eine zentrale Komponente am Körper oder in der unmittelbaren Umgebung des Patienten übertragen. Dort kann etwa aus der Kombination von EKG- und Pulswelleninformation die Pulswellenlaufzeit abgeleitet werden. Gemessene und errechnete Daten und Informationen werden über ein Kommunikationsnetz an einen Server weitergeleitet, der diese auswertet und bei einem Notfall je nach Dringlichkeit die Verbindung zu einem Dienstleister oder Arzt aufnimmt. Für die Akzeptanz solcher Systeme ist es unerlässlich, dass – etwa durch Verschlüsselung und Signierung der übertragenen Daten – der unberechtigte Zugriff auf personenbezogene Informationen zuverlässig verhindert wird.

Bewohnern zukünftiger Wohnungen wird ein Angebot verschiedener Dienstleistungen aus unterschiedlichen Bereichen der Informations- und Kommunikationstechnik zur Verfügung gestellt werden. Durch deren Nutzung lässt sich das Leben in den eigenen vier Wänden komfortabler und sicherer gestalten. Ein „Smart Living Manager“ ermöglicht den Bewohnern die Verwaltung, Konfiguration und Nutzung dieser Dienste über das gewohnte eigene TV-Gerät. Sie können jeweils einzeln zur Verfügung gestellt werden, ihre Nutzung setzt keine aufwendige Hausinfrastruktur voraus. Dies gilt grundsätzlich auch für Aktoren oder Sensoren, für die jedoch entsprechende Ausbaustufen dienstabhängig vorgesehen werden.

Automatisches oder manuelles Zusammenstellen von Informationen verschiedener Anbieter erfordert kein Eingreifen des Bewohners, sondern erfolgt für diesen transparent entsprechend der individuellen Konfiguration. Das Bereitstellen des Contents erfolgt sowohl durch Push- als auch durch Pulltechniken. Durch einen Kalenderservice erhält der Bewohner die Möglichkeit, die Kommunikation mit verschiedenen Dienstleistern abzuwickeln, Verfügbarkeiten zu prüfen, Termine einzutragen bzw. zu buchen. Dabei werden für alle Angebote mindestens Preis und Dauer der Dienstleistung angezeigt. Darüber hinaus erleichtern Zusatz- bzw. Hintergrundinformationen Auswahl und Verständnis. Ein Lieferservice richtet sich an alle Bewohner, die aufgrund unterschiedlicher Bedingungen (Kinderbetreuung, Beruf, Erkrankung, eingeschränkte Mobilität) nicht in der Lage sind, Waren des täglichen Bedarfs vor Ort persönlich auszuwählen und zu kaufen. Anbieter aus der Region ermöglichen nicht nur aktuelle Angebote für Waren des täglichen Bedarfs, sondern können aufgrund der geringen Entfernung des Bewohners zum Lieferanten eine schnelle Lieferung gewährleisten. Darüber hinaus entfällt weitgehend der Aufwand für die Auswahl von Waren im Ladenlokal und deren Transport. Über Community-Services (soziale Nachbarschaftsdienste) können die Bewohner eines Quartiers miteinander kommunizieren. So können innerhalb der „sozialen Nachbarschaft“ Aktivitäten oder Hilfestellungen bei nachbarschaftlichen Arbeiten organisiert werden.

It is indispensable for the acceptance of such systems that unauthorised access to personal information is reliably prevented – through measures such as encoding or encryption of the transferred data.

Residents of future dwellings will be provided with a variety of services from different areas of information and communication technology. Using these will make living in their own home more comfortable and safer. A “Smart Living Manager” will enable residents to manage, configure and use these services via their own, familiar television set. They can be made available as individual sets; their use does not require costly building infrastructure. This generally also applies to actuators and sensors, which, however, are subject to expansion stages depending on the respective services.

Automatic or manual compilation of information from different providers requires no active part on behalf of the resident, but is carried out transparently based on the resident's individual configuration. Content is made available via push as well as pull techniques. A calendar service gives residents the possibility to communicate with different service providers, check availability, and enter or book appointments. In this process all offers are displayed with at least the price and the duration of the service. Furthermore, additional information makes choosing and understanding easier. A delivery service is available for all residents who due to a range of conditions (child care, job, illness, mobility) are not able to personally choose and buy goods for everyday use in the shops. Regional suppliers not only provide daily goods, but due to the small distance between resident and supplier they can ensure quick delivery. Furthermore, the costs of renting shop premises and transporting goods can mostly be saved. Through community services (neighbourhood services) residents of a dwelling can communicate with one another. Thus within the “social neighbourhood” activities can be organised or help can be provided for neighbourly activities.

3. Hurdles for Ambient Assisted Living

Similar to lives, aging processes are also highly individual, which is reflected in a great variety of individual needs, lifestyles and diseases (“Individual Aging”). Admittedly, comprehensive AAL concepts, which will sufficiently cover all potentially arising needs, are in principle imaginable in some areas. However, these solutions are hardly practical and, above all, not affordable for the majority of the population. Affordable AAL solutions, which are aimed at the whole population, must be modular and extendable, and thus cost-efficiently adjustable to changing individual needs. Therefore there are no “one size fits all” solutions, at least not in the foreseeable future. On the other hand, in order to combine individual specific solutions, existing gaps between individual systems often have to be bridged by routine activities carried out by human personnel – resources which could better be invested in the individual care of the end user.

3. Hürden für Ambient Assisted Living

Wie Lebensläufe sind auch Alterungsprozesse hochgradig individuell, was sich in einer Vielzahl von individuellen Bedürfnissen, Lebensstilen und Krankheitsentwicklungen manifestiert („Individual Aging“). Zwar sind in einigen Bereichen prinzipiell umfassende AAL-Konzepte denkbar, die alle im Laufe der Zeit potenziell auftretenden Bedürfnisse hinreichend abdecken. Entsprechende Lösungen sind allerdings kaum praktikabel und vor allem für den größten Teil der Bevölkerung nicht finanzierbar. Bezahlbare AAL-Lösungen, die sich an die Gesamtheit der Bevölkerung richten, müssen modular und erweiterbar sein und sich damit kosteneffizient an sich verändernde individuelle Bedürfnisse anpassen lassen. Es gibt also zumindest auf absehbare Zeit keine Perspektive für „one size fits all“-Lösungen. Um andererseits einzelne partikulare Lösungen miteinander zu kombinieren, müssen bestehende Lücken zwischen Einzelsystemen häufig durch Routinetätigkeiten menschlichen Personals überbrückt werden – Ressourcen, die besser in die individuelle Betreuung der Endnutzer investiert werden könnten.

Je nach individuellem Bedarfsprofil umfassen die erforderlichen AAL-Lösungen jeweils Teilsysteme aus unterschiedlichen Bereichen, z. B. Gebäudeautomation, Haussicherheit, Telemonitoring, ambulante Medizin etc. Aufgrund mangelnder Motivation der Anbieter, aber insbesondere wegen fehlender bzw. inkompatibler Schnittstellenstandards, Datenaustauschprotokolle und Datenmodelle werden Teilsysteme derzeit praktisch nicht bereichsübergreifend gekoppelt, wodurch erhebliche Synergiepotenziale unerschlossen bleiben. So könnten z. B. Informationen über Status und Trends der Benutzersituation, z. B. eine nicht erfolgte Medikamenteneinnahme, aber auch verfügbare Ressourcen wie Dienste und Aktuatoren von den Teilsystemen sinnvoll gemeinsam genutzt werden. Ebenso ist es bislang unmöglich, höherwertige Funktionalitäten durch Kombination aufeinander aufsetzender Einzelfunktionen zu generieren. Dadurch kann ein erheblicher Anteil potenziell möglicher Unterstützungsleistungen nicht oder nur suboptimal erbracht werden.

Die geschilderten Integrationsdefizite sind umso erstaunlicher, als innerhalb zahlreicher derzeit entwickelter oder bereits betriebener AAL-Anwendungen, z. B. in Hausnotruf-, Sicherheits- oder Telemonitoring-Systemen, durchaus vergleichbare Hardware- und Softwarekomponenten zum Einsatz kommen, die für verschiedene Funktionen bzw. Einzelanwendungen gemeinsam genutzt werden könnten, um Aufwand und Kosten zu sparen. Diese Einsparungsmöglichkeit wird von Endkunden, Dienstleistern und Systemintegratoren zwar erkannt, doch wird die gemeinsame Nutzung von Ressourcen und Diensten unterschiedlicher Hersteller aufgrund von fehlenden Anreizen für die Hersteller, aber auch regulatorischen Anforderungen an die Verlässlichkeit (Verfügbarkeit, Sicherheit, Zuverlässigkeit) der Systeme, insbesondere im Medizin- und Sicherheitsbereich, derzeit nicht unterstützt. Um diese Hürde zu überwinden, sind entsprechende Geschäftsmodelle, Standards, Zertifizierungsprogramme und Qualitätssicherungsmechanismen erforderlich.

According to the individual needs profile, the required solutions consist of subsystems from different areas such as building management systems, building security, telemonitoring, outpatient medicine, etc. Due to a lack of motivation on behalf of the providers, but particularly due to a lack of missing or incompatible interface standards, data exchange protocols and data models, subsystems from different areas are currently not linked, thus leaving significant synergy potential untapped. Information about status and trends of the user's situation, such as not having taken their medication, but also available resources such as services and actuators, could be used jointly by the subsystems. Similarly, it has not yet been possible to generate higher quality functionality by combining individual functions that are based on one another. Therefore a significant part of potentially possible support services can either be provided only in a suboptimal way or not at all.

The integration deficits described above are all the more astonishing since several applications that are currently being developed or already being used, for example in home alert, security, and telemonitoring systems, use quite comparable hardware and software components, which could be used jointly for different functions or individual applications in order to save effort as well as costs. Even though this opportunity to save is recognised by end customers, service providers and system integrators, the joint use of resources and services by different manufacturers is currently not supported due to a lack of incentives for manufacturers, but also due to regulatory demands on the systems' reliability (availability, safety, reliability), particularly in the medical and security sector. In order to overcome this hurdle, appropriate business models, standards, certification programmes and quality assurance mechanisms are required.

For a successful and sustainable implementation of the AAL concept it is absolutely essential to involve the complete chain of parties in the health care sector, in order to develop new forms of collaborations between everyone involved and to be able to optimise the existing process chains. Identifiable hurdles on the way are rigid structures, specific individual interests and distributed cost and effect relationships – whoever covers the cost does not benefit from potential savings. New, overarching business models that comply with the health system are required in order to guarantee balanced improvements in medical care with regards to quality and efficiency.

A great challenge for the flexibility of assistive systems is the more and more dominating phenomenon that with increasing age the number of diseases per individual is rising – every third person over 70 has at least five different diseases and thus a clearly increased risk of being significantly functionally impaired. Such restrictions influence each other in such a negative way that solutions which have been working well for a longer period of time (such as controlling insulin requirement) can suddenly, after another impairment (e. g., of the visual apparatus), no longer be used. This often requires replacing established solutions or at least adjusting the appropriate

Zur erfolgreichen und nachhaltigen Umsetzung des AAL-Konzepts ist es unbedingt erforderlich, alle Beteiligten der Gesundheitsversorgung einzubeziehen, um neue Formen der Zusammenarbeit zu entwickeln und die relevanten Prozessketten optimieren zu können. Als Hürden auf diesem Weg lassen sich starre Strukturen, Partikularinteressen und verteilte Kosten- und Wirkungsbeziehungen – der Kostenträger profitiert nicht selbst von potenziellen Einsparungen – identifizieren. Es sind neue übergreifende gesundheits-systemkonforme Geschäftsmodelle erforderlich, um ausgewogene Verbesserungen der medizinischen Versorgung hinsichtlich Qualität und Effizienz sicherzustellen.

Eine große Herausforderung an die Flexibilität assistiver Systeme stellt das mit zunehmendem Alter immer dominierender werdende Phänomen der erhöhten Krankheitshäufigkeit dar – jeder dritte über 70-Jährige hat mindestens fünf verschiedene Erkrankungen und somit ein deutlich erhöhtes Risiko, funktionell erheblich beeinträchtigt zu sein. Entsprechende Einschränkungen beeinflussen sich dabei in derart negativer Weise, dass Lösungen, die längere Zeit sehr gut funktioniert haben – wie z. B. die Regelung des eigenen Insulinbedarfs – nach einer weiteren Einschränkung – etwa des Sehapparates – plötzlich nicht mehr eingesetzt werden können. Dies erfordert häufig den Austausch der bereits etablierten Lösungen oder zumindest die Anpassung der entsprechenden Bedienschnittstellen. Da derartige Anpassungen primär durch die sich verringernden sensorischen, kognitiven und motorischen Fähigkeiten der Endnutzer getrieben werden, sollten nach Möglichkeit alle betroffenen Bedienschnittstellen der verwendeten AAL-Lösungen in einheitlicher Weise angepasst werden können.

4. Interoperabilität für Ambient Assisted Living

Aus den geschilderten Randbedingungen ergibt sich, dass sich Art und Weise der Entwicklung von Komponenten und Plattformen für zukünftige AAL-Anwendungen sehr stark vom klassischen „Single-Box“-Design unterscheiden: Das erforderliche Zusammenwirken vieler, überdies wechselnder Komponenten und Geräte legt nahe, dass sich diese dynamisch zu Ensembles zusammenfinden können.[23] Mit der steigenden Zahl an der Bearbeitung einer Aufgabe beteiligter Objekte sind auch neue Strategien für das Zusammenspiel der einzelnen Komponenten notwendig.[24] Der wichtigste Aspekt hierfür ist die Interoperabilität, auf deren Basis unabhängige heterogene Einheiten zusammenarbeiten können. Wichtig dabei ist, dass dies nicht nur für Geräte eines Herstellers gilt, was letztendlich bedeutet, dass herstellerübergreifende Standards definiert werden müssen, die nicht nur Übertragungsverfahren und Syntax, sondern auch die Semantik aller Daten beschreiben, welche die Geräte austauschen.

user interfaces. Since these adjustments are primarily driven by the decreasing sensory, cognitive and motor skills of the end users, all interfaces of the applied solutions, which are affected, should be adjusted consistently.

4. Interoperability for Ambient Assisted Living

The conditions described above make it clear that the development of components and platforms for future applications will differ significantly from the classic “single box” design: the required interoperability of many (and also changing) components and devices suggests that these will have to dynamically group themselves into ensembles.[23] The growing number of objects involved in the completion of a task also requires new strategies for the interaction of individual components.[24] The most important aspect here is the interoperability, on the basis of which heterogeneous units can work together. In this context it is important that this not only applies to all devices by one manufacturer. Ultimately this means that standards which not only describe the communication method, but also the semantics of all data that are exchanged by the devices have to be defined across manufacturers.

5. Strategies and recommendations

On the basis of dynamic software infrastructure solutions, the existing individual solutions that support us in the areas of preventative health, diagnostics, therapy and care can be combined and integrated in a coherent complete solution within the complete supply and value chain. Technologies, products and services have to form holistic system solutions, which are nevertheless adjustable to individual needs, lifestyles and developments of diseases. The Fraunhofer Allianz Ambient Assisted Living has set itself the goal to develop a corresponding, flexible, modular and extendable platform and at the same time initiate standardisation processes in order to achieve its market-wide acceptance. Experts within the Fraunhofer Allianz AAL are currently documenting the state of the art on the basis of the most important relevant standards and norms.[25]

A demand for complete solutions:

The most important step in the development is to connect the individual applications and products which already exist today and combine them with the relevant players of the complete supply and value chain. However, currently sustainable, complete service concepts are not in sight. The existing, isolated applications include home alert systems which today are still designed as pure alarms, sensors for light control, and device-specific user interfaces. Today such isolated applications and products can only be combined to form complete solutions with the help of costly so-called home control systems, where each individual change requires a system specialist – this considerably increases the price of such complete solutions. Therefore funding programmes are to support projects for

5. Strategien und Handlungsempfehlungen

Auf der Basis von dynamischen Software-Infrastrukturlösungen können die existierenden Einzellösungen, die uns bei Gesundheitsvorsorge, Diagnostik, Therapie und Pflege unterstützen, zu einer stimmigen Gesamtlösung im Rahmen der kompletten Versorgungs- und Wertschöpfungskette verbunden und integriert werden. Technologien, Produkte und Dienstleistungen müssen dabei ganzheitliche Systemlösungen bilden, die jedoch an individuelle Bedürfnisse, Lebensstile und Krankheitsentwicklungen anpassbar sind. Die Fraunhofer Allianz Ambient Assisted Living hat sich zum Ziel gesetzt, eine entsprechende flexible, modulare und erweiterbare Plattform zu entwickeln und flankierend Standardisierungsprozesse zu initiieren, um deren marktbreite Durchsetzung zu ermöglichen. Derzeit dokumentieren Experten innerhalb der Fraunhofer Allianz AAL anhand der wichtigsten relevanten Standards und Normen den Stand der Technik.[25]

Forderung nach Gesamtlösungen:

Der wichtigste Entwicklungsschritt besteht darin, die heute bereits vorhandenen Einzelanwendungen und -produkte zu verknüpfen und mit den relevanten Akteuren der kompletten Versorgungs- und Wertschöpfungskette zu verbinden. Derzeit sind tragfähige durchgängige Dienstleistungskonzepte jedoch noch in keiner Weise absehbar. Zu den vorhandenen Insellösungen zählen Hausnotrufsysteme, die heute noch als reine Alarmmelder konstruiert sind, Sensoren zur Lichtsteuerung oder gerätespezifische Benutzungsschnittstellen. Solche Einzelanwendungen und -produkte können heute nur durch aufwendige so genannte Haussteuerungssysteme zu Gesamtlösungen verknüpft werden, wobei Änderungen stets erneut einen Systemspezialisten erfordern, was derartige Lösungen erheblich verteuert. Im Rahmen von Förderprogrammen sollten deshalb Projekte für Gesamtlösungen unterstützt werden, die im Sinne der oben dargestellten Plattform eine möglichst vollständige Abdeckung der Bedürfnisse und zugrunde liegenden Befindlichkeiten bzw. Krankheitsbilder eines Menschen gewährleisten.

Normierung, Zertifizierung und Standardisierung:

Für Ambient Assisted Living werden Produkte und Komponenten unterschiedlicher Hersteller aus den verschiedensten Bereichen eingesetzt. Mangels standardisierter Schnittstellen und Protokolle sind Austauschen, Kombinieren und Nachrüsten von Systemkomponenten im Sinne herstellerübergreifender Gesamtlösungen kaum möglich. Erforderlich ist ein System von Festlegungen, das über Herstellergrenzen hinweg Schnittstellen auf funktionaler und semantischer Ebene vereinheitlicht. Als Ausgangspunkt dazu können Normen bzw. Standards aus der Gebäudeautomatisierung oder dem „Personal Health"-Bereich der Medizintechnik dienen, die in diesen Bereichen schon weit verbreitet sind. Ebenso sollten Industrie und weitere relevante Akteure aus diesen und weiteren AAL-Teilbereichen nicht nur in die Normungs- bzw. Standardisierungsarbeit, sondern auch in weiterführende, darauf aufbauende Aktivitäten der Marktentwicklung – u. a. die qualifizierte Zertifizierung standardkonformer

complete solutions which, in line with the platform outlined above, provide, if possible, a complete cover for the needs and underlying problems or symptoms of the individual.

Standardisation and certification:

Products and components from different manufacturers from a wide and diverse range of areas are used for Ambient Assisted Living. Due to a lack of standardised interfaces and protocols the replacement, combination and refitting of system components as aimed at in manufacturer-independent, complete solutions are hardly possible. What is required is a system of rules that standardises interfaces on a functional and semantic level across manufacturers. Norms and standards from the building management systems sector or the personal health area of medical technology, which are already widespread in these areas, could serve as a starting point. Furthermore, the industry and other relevant players from these and other AAL sub-areas should be significantly involved not only in the standardisation work, but also in further activities of market development, such as the qualified certification of standards-compliant components, which are built on the initial standards. In this context the Continua activities could also serve as an example. All this would significantly increase the market chances and sales of such products, since networkable products can be offered for an overall cheaper price, thus providing the customer with substantial added value.

A consistent design of the supply and value chain:

Particularly from a health economic perspective, the use of AAL technologies always carries the risk of relatively high procedural costs, which might not be offset by savings resulting from people remaining in their own home for longer, a reduction of ambulatory care, a decrease in emergency responses and hospitalisation (or its duration) as well as medical outpatient care, etc. One of the main reasons for this is that health-related assistive systems currently only cover parts of the supply and interoperability chain, which stretches from the individual or system components located close to their body (sensors in particular) to the doctor's surgery, the outpatient department and hospital and the representatives of the different health professions working there. A consistent design of this chain makes it possible to on the one hand implement assistive functions in the home which will enable particularly elderly people to independently live in their own home for as long as possible, and on the other hand to combine them with case- or person-specific information, monitoring and decision support systems, which are based on the integration of health-relevant data and services for the respective person. In order to implement this concept (particularly with appropriate business models in mind) it is indispensable to involve the complete chain of players in the health care sector, and to get – besides technology developers – medical professionals, health insurances, associations, service providers in the social and health sectors, advocacy groups, representatives of the housing industry and psychologists to the table in order to develop new forms of cooperations between everyone involved. Apart from social insurance institutions, government bodies (on different levels) as well as inde-

Komponenten – maßgeblich einbezogen werden. All dies würde Marktchancen und Absatz entsprechender Produkte erheblich steigern, da vernetzungsfähige Produkte insgesamt billiger angeboten werden können und in der Hand des Kunden einen erheblichen Mehrwert bieten.

Durchgängige Gestaltung der Versorgungs- und Wertschöpfungskette:
Insbesondere aus gesundheitsökonomischer Sicht besteht beim Einsatz von AAL-Technologien prinzipiell das Risiko relativ hoher Verfahrenskosten, die möglicherweise nicht durch Einsparungen aufgrund eines längeren Verbleibens in der eigenen Wohnung, einer Reduktion des ambulanten Pflegeaufwandes, einer Verringerung von Notfalleinsätzen und Krankenhausaufenthalten (bzw. deren Verweildauern) oder ambulanter medizinischer Leistungen etc. kompensiert werden. Eine der Hauptursachen hierfür ist wiederum, dass gesundheitsbezogene Assistenzsysteme zurzeit lediglich Teilbereiche der Versorgungs- und Interoperabilitätskette abdecken, die vom Individuum bzw. von körpernahen Systemkomponenten (insbesondere Sensoren) bis zu Arztpraxis, Ambulanz oder Klinik und den darin arbeitenden Vertretern der verschiedenen Heilberufe reicht. Mit einer durchgängigen Gestaltung dieser Kette wird es möglich, einerseits häusliche Assistenzfunktionen zu realisieren, die es insbesondere älteren Menschen ermöglichen, möglichst lange selbstbestimmt in den „eigenen vier Wänden" zu leben und diese andererseits mit fall- bzw. personenbezogenen Informations-, Überwachungs- und Entscheidungsunterstützungssystemen zu kombinieren, die auf der Integration gesundheitsbezogener Daten und Dienste für die betreffende Person basieren. Zur Umsetzung dieses Konzepts (insbes. im Sinne entsprechender Geschäftsmodelle) ist es ist es unerlässlich, die vollständige Kette der Akteure in der Gesundheitsversorgung einzubeziehen und neben den Technologieentwicklern Mediziner, Krankenkassen, Sozial- und Gesundheitsdienstleister, Wohnungswirtschaft, Psychologen, Verbände und Interessenvertretungen an einen Tisch zu bekommen, um neue Kooperationsformen zwischen allen Beteiligten zu entwickeln. Neben den Sozialversicherungsträgern sind auch staatliche Körperschaften (auf verschiedenen Ebenen) und freie Wohlfahrtsverbände sowie Genossenschaften und Träger der privaten und öffentlichen Wohnungswirtschaft zu berücksichtigen, insbesondere wenn auch bauliche Veränderungen im größeren Maßstab als Teil von auf AAL-Technologien basierenden Produkten bzw. Plattformen in Frage kommen. Gleichzeitig sind neue gesundheitssystemkonforme Geschäftsmodelle zu entwickeln, die ausgewogene Verbesserungen der medizinischen Versorgung hinsichtlich Qualität und Effizienz sicherstellen.

Strategien für die Öffentlichkeitsarbeit:
Die Aufklärung und Information über Möglichkeiten der Unterstützung durch AAL-Systeme muss verstärkt werden. Bislang sind die technischen Strukturen und Funktionen bei potenziellen Anwendern viel zu wenig bekannt. Da die Akzeptanz auch vom Grad der Informiertheit abhängt, müssen Strategien für die Öffentlichkeitsarbeit erarbeitet werden. Dazu gehört auch, dass Pflegestützpunkte und

pendent welfare organisations, co-operatives and providers of the private and public housing industry have also be taken into account, particularly if structural changes on a larger scale are to be considered as part of products or platforms based on Ambient Assisted Living technologies. Simultaneously new, overarching business models that comply with the health system are required in order to guarantee balanced improvements in medical care with regards to quality and efficiency.

Strategies for public relations:
Education and information about possible support through AAL systems has to be increased. To date the technical structures and functions are not known well enough amongst potential users. Since the acceptance depends on how well informed the public is, public relation strategies have to be developed. This also includes that care centres and similar institutions have to be informed about current developments. In connection with this VDE activities have to be continued and strengthened; by involving regional associations and local disseminators as well as social and community contacts, regional potentials and specifics of the demographic developments can be used increasingly to distribute the ideas. The areas in which the problems are the strongest offer the best chances to demonstrate today's possibilities and the perspectives the use of Ambient Assisted Living technologies provide.

6. Conclusion and outlook

Due to the lack of interoperability and complete solutions, existing AAL systems have previously not been able to tap the full potential of the AAL concept since they only cover partial aspects of possible application scenarios and situations or symptoms respectively, and allow only limited combinations of subsystems or modules. Their use has an unfavourable cost-effect relation of e.g., 17,300 euros per year of life gained (in the case of heart insufficiency[26]). However, usually elderly people have several health problems; they do not just have to deal with one disease, but several at the same time. Therefore only if the whole range of symptoms of their diseases as well as their needs are covered can costs actually be saved; since only then the costs of moving into a retirement home, hospitalisation, etc., which have been saved by remaining in their own home, can be offset.

Exactly this is not possible with today's products for the implementation of individual solutions. However, living at home for longer and the resulting savings potential with good reason play a key role in the argument for corresponding research and research funding. From the point of view of the Fraunhofer Allianz AAL, there is no alternative to the recommendations presented here if the expectations in AAL are to be fulfilled and market development as well as penetration for AAL system solutions are to be facilitated.

entsprechende Einrichtungen über AAL-Entwicklungen auf dem Laufenden sein müssen. Fortzusetzen und zu verstärken sind diesbezüglich auch VDE-Aktivitäten; durch Einbeziehung der Bezirksvereine, lokaler Multiplikatoren sowie sozialer und kommunaler Ansprechpartner können verstärkt regionale Potenziale und Spezifika der demografischen Entwicklung für die Verbreitung der Ideen genutzt werden. Wo die Probleme regional am gravierendsten sind, bieten sich die besten Chancen, die heutigen Möglichkeiten und die Perspektiven des Einsatzes von Ambient-Assisted-Living-Technologien aufzuzeigen.

6. Fazit und Ausblick

Aufgrund mangelnder Interoperabilität und fehlender Gesamtlösungen konnten bisher vorhandene AAL-Systeme nicht das Potenzial des AAL-Konzepts ausschöpfen, da sie nur Teilaspekte möglicher Anwendungsszenarien und Befindlichkeiten bzw. Krankheitsbilder abdecken sowie nur eingeschränkt Kombinationen von Teilsystemen oder Modulen erlauben. Ihr Einsatz weist entsprechend ungünstige Kosten-Effekt-Relationen von z. B. 17.300 Euro pro gewonnenem Lebensjahr (bei Herz-Insuffizienz[26]) aus. Üblicherweise addieren sich aber bei älteren Menschen mehrere gesundheitliche Probleme; sie haben nicht nur mit einer, sondern mit mehreren gleichzeitigen Krankheiten umzugehen. Entsprechend können nur bei Abdeckung des gesamten Befindlichkeits- und Krankheitsbildes tatsächlich Kosten eingespart werden, da nur dann auch der durch den Verbleib in der eigenen Wohnung eingesparte Aufwand für Heimunterbringung, Klinikaufenthalte etc. kostenwirksam wird.

Mit den heutigen Produkten zur Realisierung von Einzellösungen ist eben dies nicht möglich. Dennoch – und zu Recht – stehen längeres Leben zu Hause und die damit verbundenen Sparpotenziale im Mittelpunkt der Argumentation zugunsten entsprechender Forschung bzw. Forschungsförderung. Die vorliegenden Handlungsempfehlungen sind aus Sicht der Fraunhofer Allianz AAL alternativlos, wenn die an AAL geweckten Erwartungen eingelöst und Marktentwicklung und Marktdurchdringung für AAL-Systemlösungen ermöglicht werden sollen.

REHACARE INTERNATIONAL
DIE MESSE THE TRADE FAIR

Internationale Leistungsschau der Rehabilitations- und Pflegebranchen
International rehabilitation and care industry showcase

Die internationale Fachmesse REHACARE am Messeplatz Düsseldorf bietet behinderten, pflegebedürftigen und chronisch kranken Menschen sowie Fachleuten aus Rehabilitation, Pflege, Hilfsmittelindustrie und Sanitätsfachhandel alljährlich die Chance, sich über innovative Rehabilitationstechniken, aktuelle Pflegekonzepte und Therapien zu informieren.

Die REHACARE wurde 1977 auf Initiative der Betroffenenverbände und zahlreicher Sozialleistungsträger aus der Taufe gehoben. Ziel war die Schaffung von Chancengleichheit für Menschen mit Behinderung sowie deren Eingliederung in die Gesellschaft durch eine verbesserte Hilfsmittelversorgung.

Heute ist die REHACARE die weltweit führende Fachmesse der Rehabilitations- und Pflegebranchen mit rund 800 Ausstellern aus mehr als 30 Ländern. Das Angebot erstreckt sich über sechs Messehallen und umfasst Lösungen für alle Lebensbereiche und Zielgruppen. Vor allem im Bereich Mobilität bietet die REHACARE eine breite, hochwertige Produktpalette mit Rollstühlen, Fahrgeräten, Gehhilfen, angepassten Kraftfahrzeugen und den dazugehörigen Dienstleistungen.

Rund 120 Organisationen und Verbände für behinderte und pflegebedürftige Menschen unterstützen die Veranstaltung fachlich. Sie spielen eine führende Rolle bei der Ausrichtung von Themenparks und Sonderschauen, die für das Messekonzept charakteristisch sind. Sie greifen aktuelle Themenstellungen auf in Bereichen wie Bildung und Beruf, Behindertensport, barrierefreie Kommunikation und Architektur, Reiseangebote für Menschen mit Behinderung sowie Kunst und Kultur.

REHACARE, the international trade fair staged at the fairgrounds in Düsseldorf, provides people with disabilities, requiring care and the chronically ill as well as specialists active in rehabilitation, care, the assistive devices industry and the sale of medical supplies an annual opportunity to catch up on the latest in innovative rehabilitation technology, care concepts and therapies.

Associations representing people affected by disabilities together with numerous social services organisations initiated the drive to establish REHACARE back in 1977. The objective was to provide equal opportunities for people with disabilities and further their integration into society through improved distribution of assistive devices.

Today, as the world's No. 1 trade fair for the rehabilitation and care industry, REHACARE features some 800 exhibitors from more than 30 countries. The spectrum of offerings spread across six halls encompasses solutions for all aspects of life and all target groups. REHACARE presents an especially broad showcase of high-quality mobility products including wheelchairs, vehicles, walking aids, adapted cars and all the related services.

About 120 organisations and associations that represent people with disabilities or requiring care support the event with their expertise. They play a key role in shaping the event's trademark theme parks and special presentations. These highlights tackle topical issues such as training and careers, sport for the disabled, barrier-free communications and architecture, travel offers for people with disabilities as well as art and culture.

Die REHACARE gilt als wesentlicher Impulsgeber für die gesellschaftliche Integration von Menschen mit Behinderung, Pflegebedarf und chronischen Krankheiten, als Sprachrohr der Betroffenen und ihrer Interessenvertretungen.

Das Thema Design spielt bei der REHACARE traditionell eine wichtige Rolle. Bereits 1987 fand in Düsseldorf ein Symposium unter dem Titel „Design für Alle“ statt, das Produkte forderte, die in Funktion und Gestaltung für alle Menschen geeignet sind.

2004 wählte die Europäische Kommission die REHACARE als Plattform für die erste europäische Verleihung des „Design for All and Assistive Technology Award“.

Mit dem REHACARE DESIGN AWARD, der in Kooperation mit red dot projects ausgelobt wurde, bot die Messe Düsseldorf den Ausstellern der REHACARE die Chance, sich an einem hochkarätigen internationalen Designwettbewerb zu beteiligen.

REHACARE has earned a reputation for spurring on the integration of people with disabilities, requiring care or chronic illnesses into the community as well as acting as a mouthpiece for such groups and their interests.

Design has traditionally been a central concern at REHACARE. As far back as 1987, a symposium bannered “Design for All”, which called for products whose form and functions were suitable for all people, was held in Düsseldorf.

In 2004, the European Commission selected REHACARE as the venue for the first European “Design for All and Assistive Technology Award” prize-giving ceremony.

With the REHACARE DESIGN AWARD, which is conferred in co-operation with red dot projects, Messe Düsseldorf offered its exhibitors the opportunity to participate in a high-calibre international design competition.

REHACARE DESIGN AWARD
DER WETTBEWERB THE COMPETITION

Der Wettbewerb
The competition

REHACARE
DESIGN AWARD
2008/2009

Eine hohe Lebensqualität und Teilhabe sind für Menschen mit Handicap nicht selbstverständlich. Um dies zu verwirklichen, benötigen sie maßgeschneiderte Produkte, die eine verlässliche Funktionalität bieten, sich einfach in den Alltag integrieren und intuitiv handhaben lassen. Neben kranken und behinderten Menschen erlangt der wachsende Markt der älteren und pflegebedürftigen Menschen einen immer höheren Stellenwert, wodurch sich für die Hersteller von Rehaprodukten und Hilfsmitteln neue Herausforderungen ergeben. Zudem führen die veränderten Ansprüche der Anwender dazu, dass das Bewusstsein für designorientierte Produkte den Weg in die Branche gefunden hat und immer weiter zunimmt. Die Unternehmen können hier Maßstäbe setzen, wenn Innovationen in diesem hochsensiblen Bereich neben ihrer ausgereiften Funktionalität auch eine emotionale Qualität aufweisen, die ihre Produkte in der täglichen Anwendung selbstverständlich macht.

Der REHACARE DESIGN AWARD 2008/2009 greift diese Entwicklung in der Branche auf und zeichnet designorientierte Produkte der Reha- und Hilfsmittelbranche aus, die in innovativer Weise den Anforderungen, die ein sensibles und anspruchsvolles Umfeld an sie stellt, entsprechen.

Der erstmals durchgeführte Wettbewerb wurde von red dot projects in Kooperation mit der Messe Düsseldorf ins Leben gerufen. Eine vierköpfige internationale Jury aus anerkannten Designexperten beurteilte die eingereichten Produkte hinsichtlich ihres Innovationsgrades, ihrer Funktionalität, formalen Qualität, Ergonomie, Langlebigkeit, Produktperipherie, Selbsterklärungsqualität und ökologischen Verträglichkeit.

Ausgezeichnet wurden insgesamt 24 Produkte, von denen vier mit einer besonderen Anerkennung gewürdigt wurden. 20 Produkten wurde die Auszeichnung REHACARE DESIGN AWARD: GOOD DESIGN zuteil. Drei Produkte stachen durch eine besondere gestal-

A high quality of life and playing an active role in society is not a given for handicapped people. In order to achieve these, they require tailor-made products that offer reliable functionality, easily integrate into daily life, and can be used intuitively. Besides sick and disabled people, the growing market of older people and people requiring care plays an increasingly important role, which results in new challenges for the manufacturers of rehabilitation products and aids. In addition, design awareness has found its way into the industry due to changed user demands and is growing continuously. In this area companies can set standards if innovations in this highly sensitive area – in addition to their sophisticated functionality – also feature an emotional quality which makes the daily use of the products a matter of course.

The REHACARE DESIGN AWARD 2008/2009 takes up this development in the industry and pays tribute to design-oriented products from the rehabilitation and aid industry which innovatively meet the requirements placed on them by a sensitive and demanding environment.

The competition, which was held for the first time, has been initiated by red dot projects in collaboration with Messe Düsseldorf. A jury of four, consisting of recognised design experts, assessed the submitted products with regards to their degree of innovation, functionality, formal quality, ergonomics, durability, product periphery, self-explanatory quality, and ecological compatibility.

In total 24 products won an award; four of them received special recognition. 20 products received the REHACARE DESIGN AWARD: GOOD DESIGN. Three products stood out due to their particularly innovative design and were therefore awarded the REHACARE DESIGN AWARD: BEST DESIGN. The entry with the highest innovative strength received the distinction REHACARE DESIGN AWARD: BEST INNOVATION.

terische und innovative Umsetzung hervor, sodass sie die Auszeichnung REHACARE DESIGN AWARD: BEST DESIGN erhielten. Die Einreichung mit der höchsten Innovationskraft erhielt die Auszeichnung REHACARE DESIGN AWARD: BEST INNOVATION.

Mit dem REHACARE DESIGN AWARD setzen die Initiatoren Impulse, um das höhere Bewusstsein für gut gestaltete und hochfunktionale Produkte in der Reha- und Hilfsmittelbranche zu würdigen und weiter voranzutreiben. Indem sich die Unternehmen ihren Mitbewerbern im internationalen Vergleich stellen, bietet ihnen der Wettbewerb die Möglichkeit zur Positionierung. Die Auszeichnung selbst verschafft den Verbrauchern Orientierung, denn sie dokumentiert die hohe gestalterische Qualität des jeweiligen prämierten Produktes.

Die Vielzahl der Einreichungen zu diesem Wettbewerb zeigt eindrucksvoll, dass schon jetzt innovative Designlösungen verwirklicht werden, die weit über rein technische Ansätze hinausgehen. Die Qualität der Produkte ist nicht zuletzt durch den Einsatz neuester Materialien so hoch wie nie zuvor, das negativ konnotierte Medizinische nimmt sich zurück und Aspekte der Haptik und Hygiene bestimmen die Designentwürfe. So zeigt der Wettbewerb nicht nur herausragende Produkte der Branche, sondern erweist sich als Gradmesser für das sich wandelnde Designbewusstsein in dieser Branche.

With the REHACARE DESIGN AWARD the initiators set impulses in order to recognise the raised awareness for well designed and highly functional products in the rehabilitation and aid industry and advance it further. Facing their competitors in this international comparison, the competition offers companies the opportunity to position themselves. The award itself provides consumers with guidance because it documents the high design quality of each award-winning product.

The large number of entries to this competition shows impressively that innovative design solutions that go far beyond purely technical approaches are already being implemented. The product quality is as high as never before, not least thanks to the use of the latest materials; negative connotations of medical aspects are lessened with haptic and hygienic aspects defining the design concepts. Thus the competition not only showcases excellent products in the industry but also turns into a benchmark for the changing design awareness in this industry.

Die Jurierung
The adjudication

Die Qualität eines Wettbewerbs hängt in hohem Maße von der Zusammensetzung, Kompetenz und Unabhängigkeit seiner Jury ab. Für den erstmals ausgeschriebenen REHACARE DESIGN AWARD 2008/2009 bestand das vierköpfige Jurygremium aus international anerkannten Designexperten, die über eine langjährige Erfahrung im Bereich der Medizin- und Rehaprodukte verfügen. Ihnen oblag es, alle angemeldeten Produkte auf die anspruchsvollen Kriterien des Wettbewerbs zu begutachten.

Bevor jedoch die eigentliche Jurysitzung im September 2008 in Düsseldorf stattfinden konnte, war der TÜV SÜD aufgerufen, eine sachliche Gebrauchstauglichkeitsprüfung auf offensichtliche technische Mängel der im Original eingereichten Produkte durchzuführen. Denn erst durch diese konnten die Arbeiten ihre Zulassung zur Jurierung erhalten.

The quality of a competition depends to a high degree on the composition, competence, and independence of its jury. The jury of four of the REHACARE DESIGN AWARD 2008/2009, which was held for the first time, consisted of internationally recognised design experts with many years of experience in the field of medical and rehabilitation products. It was their task to assess all submitted products on the basis of the competition's demanding criteria.

However, before the actual jury session could take place in Düsseldorf, Germany, in September 2008, TÜV SÜD, an international testing, evaluation and certification organisation, was asked to carry out an objective usability test and check the products, which were submitted as originals, for obvious technical deficiencies. Only after successfully passing this examination products were then admitted to the adjudication.

Während der Jurysitzung wurde deutlich, dass sich das Designbewusstsein der Branche gewandelt hat. Nicht zuletzt führen die aufgrund der demographischen Entwicklung veränderten Märkte in Verbindung mit dem hohen Innovationsgeist der Sparte zu einer härteren Marktsituation. Und gerade dort, wo ein starker Wettbewerb und ein hoher Qualitätsanspruch zusammenstoßen, erhält Design als entscheidendes Differenzierungsmerkmal eine hohe Bedeutung. Diese Erkenntnis ist bereits in der Branche angekommen. Gerade dies machte es der Jury jedoch nicht einfach, die Gewinner aus den zahlreichen hochwertigen Einreichungen auszuwählen.

Sowohl in gestalterischer wie funktionaler Hinsicht seien hervorragende Produkte zu finden, konstatierte Juror Dirk Schumann. Besonders Hilfsmittel, die durch ihren hohen emotionalen Gehalt einen integrativen Charakter besitzen, punkteten beim Gremium. Dies sei genau der richtige Ansatz, wie eine sensible Produktgestaltung zu mehr Akzeptanz und Lebensqualität der Nutzer führen kann.

Insgesamt zog die Jury eine überaus positive Bilanz des Wettbewerbs. Übereinstimmend begrüßten die Jurymitglieder die stärker werdende Entwicklung ganzheitlicher Ideen, damit funktional hochwertige Einzellösungen durch ein gestalterisch durchdachtes Konzept eine Interaktion zwischen dem Produkt, dem Nutzer und seiner Umwelt schaffen.

During the jury session it became clear that the industry's design awareness had changed. The fact that the markets have changed as a result of the demographic development, combined with the strong innovative spirit of the industry, will also lead to a tougher situation in the market. And especially where strong competition and high quality standards collide, design has great importance as a decisive differentiating feature. This insight has already reached the industry. It was exactly this fact, however, which did not make it easy for the jurors to select the winners from the numerous high-quality entries.

With regards to both design and function there were excellent products among the entries, stated juror Dirk Schumann. Especially aids which had an integrative character thanks to their high emotional content scored points with the jury. This was exactly the right approach showing how sensitive product design can lead to increased acceptance and a higher quality of life for the user.

Overall, the jury gave an extremely positive summary of the competition. The jury members unanimously welcomed the increasing development of holistic ideas enabling functionally high-end detail solutions to create an interaction between the product, the user, and their environment thanks to cleverly thought-out design concepts.

Die Preisverleihung
The award ceremony

Am 15. Oktober 2008 fand der REHACARE DESIGN AWARD 2008/2009 seinen glanzvollen Höhepunkt. Zur Preisverleihung am Eröffnungstag der 19. internationalen Fachmesse REHACARE in Düsseldorf waren neben den Gewinnern auch zahlreiche geladene Gäste erschienen, um der Ehrung der Preisträger beizuwohnen.

In seiner Einführung unterstrich Prof. Dr. Peter Zec, Jurymitglied und Initiator des red dot design award sowie Senator des internationalen Designverbandes Icsid (International Council of Societies of Industrial Design), die Wichtigkeit eines Designwettbewerbs in dieser Branche. Für ein großes Interesse an innovativen und gut gestalteten Lösungen sprächen die bereits am Eröffnungstag gestiegenen Besucherzahlen der REHACARE INTERNATIONAL, wodurch die Aktualität des Themas mehr als deutlich werde. Insbesondere die demografische Entwicklung stärke das Interesse der Menschen an hoch-

On 15 October 2008, the REHACARE DESIGN AWARD 2008/2009 celebrated a highpoint. At the award ceremony, held on opening day of the 19th annual REHACARE trade fair in Düsseldorf, winners and invited guests gathered to attend the honoring of the award winners.

In his introduction, Prof. Dr. Peter Zec, jury member and initiator of the red dot design award and Senator of the international design organization Icsid (International Council of Societies of Industrial Design), emphasized the importance of design awards in this field. The vast number of visitors to REHACARE INTERNATIONAL, on opening day alone, pointed to a strong interest in innovation and well-designed solutions, whereby the timeliness of the subject was clearly proven. Changing demographics, in particular, will increase people's interest in high-quality materials for rehabilitation and nursing care. As a result, the demands on manufacturers will continue to

wertigen Hilfsmitteln für Rehabilitation und Pflege. Damit würden auch die Anforderungen für die Hersteller in den kommenden Jahren kontinuierlich steigen. Der REHACARE DESIGN AWARD fungiert dabei nicht nur als Gradmesser für die gesamte Entwicklung der Branche, er ist ebenso ein wichtiges Orientierungsinstrument und ein neuer Antrieb für Innovationen.

Wilhelm Niedergöker, Geschäftsführer der Messe Düsseldorf, und Prof. Dr. Zec gratulierten den Gewinnern und überreichten gemeinsam die Urkunden. „Alle Preisträger können zu Recht stolz auf ihre Auszeichnung sein. Sie ist ein wichtiger Indikator für die hervorragende Qualität ihrer Produkte – eine Orientierungshilfe für Kunden und ein Positionierungskennzeichen gegenüber der Konkurrenz", resümierte Prof. Dr. Peter Zec.

Zum Ende der Veranstaltung eröffnete er die Ausstellung der Gewinner und lud die Anwesenden ein, sich selbst von dem hohen gestalterischen und funktionalen Niveau der prämierten Produkte zu überzeugen.

increase over the coming years. The REHACARE DESIGN AWARD functions in that matter not only as an indicator of overall development in the field, but also as an important benchmarking instrument and stimulus for innovation.

Wilhelm Niedergöker, Managing Director of Messe Düsseldorf, and Prof. Dr. Peter Zec together congratulated the winners and handed out the certificates. "All of the winners should be very proud of their awards. Having won one is an important indicator of a product's outstanding quality – a benchmark for customers and a mark of distinction among the competition," said Prof. Dr. Peter Zec.

To close the event, Prof. Dr. Peter Zec opened the winners' exhibition and invited the guests to see for themselves the highly creative and functional level achieved by these premier products.

Die Ausstellung
The exhibition

Auf der REHACARE INTERNATIONAL, dem internationalen Marktplatz für den Bereich Rehabilitation und Pflege, bot die erste Ausstellung zum REHACARE DESIGN AWARD vom 15. bis 18. Oktober 2008 einen faszinierenden Querschnitt der aktuellen Designentwicklungen und Trends der Branche.

Angefangen bei Inkontinenzprodukten über Mobilitätshilfen bis hin zu Produkten für Kinder mit Handicap: Das Fachpublikum wie auch interessierte Besucher konnten sich auf der Präsentationsplattform der Gewinnerprodukte von der Vielfalt dessen überzeugen, was Design heute in der Rehabilitations- und Pflegebranche schon leisten kann – sei es ein zurückhaltend gestaltetes Hilfsmittel, das sich unauffällig in die Umgebung integriert, oder ein selbstbewusst gestaltetes Produkt, mit dem sein Benutzer seine Individualität, Agilität und Lebensfreude ausdrücken kann.

Gute Gestaltung verbunden mit einem hohen Maß an Innovation und Funktionalität ist heute ein elementares Kriterium für den Erfolg von Produkten. Die lohnende Investition in Design und eine hochwertige Produktgestaltung und die damit zu erreichenden Marktvorteile werden zunehmend von den Unternehmen erkannt. Gute Ideen und durchdachte gestalterische Lösungen ebnen den Weg in eine viel versprechende Zukunft. Ebendies veranschaulichte auch die hochwertige Ausstellung der im Rahmen des REHACARE DESIGN AWARD ausgezeichneten Einreichungen.

At the REHACARE INTERNATIONAL, the international market place for the field of rehabilitation and care, the first REHACARE DESIGN AWARD exhibition from 15 to 18 October 2008 offered a fascinating cross section of current design developments and trends in the industry.

From incontinence products to mobility aids to products for children with disabilities – on the presentation platform of the winning products, the trade audience as well as interested visitors were able to satisfy themselves of the wide variety of helpful solutions which design can provide in the rehabilitation and care industry already today – be it an aid with a discreet design, which unobtrusively blends into its environment, or a product featuring a self-confident design with which users can express their individuality, agility and joy of living.

Good design in combination with a high degree of innovation and functionality is a fundamental criterion for product success today. The profitable investment in design and high-quality product design and the market advantages thus achievable are increasingly recognised by companies. Good ideas and well thought-out design solutions pave the way for a promising future. This was also illustrated by the high-quality exhibition of the entries that received a distinction in the REHACARE DESIGN AWARD.

BEST INNOVATION
GEWINNER WINNER

Die Auszeichnung REHACARE DESIGN AWARD: BEST INNOVATION wird einem Produkt verliehen, das sich durch einen besonders hohen Innovationsgrad und eine herausragende Designqualität hervorhebt.

The distinction REHACARE DESIGN AWARD: BEST INNOVATION is awarded to a product that stand out by a particularly high degree of innovation and excellent design quality.

REHACARE
DESIGN AWARD
2008/2009
BEST INNOVATION

BEST INNOVATION GEWINNER WINNER	Produkt Product	e-motion M15, 2008 Restkraftunterstützender Greifreifenantrieb Power assist for wheelchairs
	Hersteller Manufacturer	Ulrich Alber GmbH, Albstadt
	Design	einmaleins Büro für Gestaltung, Burgrieden
	Website Hersteller Manufacturer	www.alber.de
	Website Design	www.einmaleins.net

e-motion M15

Der Zusatzantrieb für manuelle Rollstühle unterstützt über in Radnaben integrierte Elektromotoren die Anschubbewegung des Rollstuhlfahrers. Das Produkt ist formal sehr zurückhaltend und reduziert gestaltet, sodass es sich wie selbstverständlich in die Optik des Rollstuhls integrieren lässt. Dank seiner unauffälligen Farbgebung eignet es sich auch für die Verwendung an optisch auffälligen Rollstühlen, ohne deren Gesamteindruck zu stören. Um einen diskreten Gebrauch zu ermöglichen, erfolgt der Antrieb ohne ein Getriebe und damit ohne eine störende Geräuschentwicklung. Menschen, die über eine eingeschränkte Kraft verfügen, erfahren mit seiner Unterstützung eine erweiterte Mobilität.

Er reagiert auf individuelle körperliche Fähigkeiten und übersetzt diese in eine angepasste Kraftunterstützung. Dadurch liefert er auch in unterschiedlichen Fahrsituationen die jeweils angemessene Hilfestellung. So werden einerseits Strecken, die bislang aufgrund ihrer Neigung nicht befahrbar waren, nun zugänglich. Andererseits ermöglicht sein Einsatz durch den geschonten Kraftaufwand des Nutzers eine Verlängerung der Fahrdauer.

Neben seinem erlebbaren funktionalen Nutzen eignet sich das Antriebsrad auch gut für Transportsituationen, da sich das gesamte Rad mit einem Knopfdruck lösen und einfach vom Rollstuhl abnehmen lässt. Das moderne Antriebskonzept trägt ebenfalls dem Aspekt der Nachhaltigkeit aufgrund seines Elektromotors und der langlebigen Lithium-Ionen-Akkus Rechnung.

This power add-on drive for manual wheelchairs uses electric motors integrated in the wheel hubs to assist the arm power of the wheelchair user. The product has a very unobtrusive and simple design so that it can be effortlessly integrated into the look of the wheelchair. Thanks to its discreet colour scheme it is also suitable for use on wheelchairs with an eye-catching design without interfering with the overall impression. In order to allow discreet use without any annoying noise, the drive does not feature a gearbox.

The drive provides people who have limited physical strength with increased mobility since it reacts to individual physical abilities, translating them into corresponding power assistance. Thus it also provides appropriate support in any driving situation, so places that have been inaccessible to wheelchairs due to their slopes can now become wheelchair-friendly. In addition to that, it enables users to cover longer distances because it allows them to save their energy.

Apart from its functional benefits while driving the wheelchair, the add-on drive is also well-suited for transport purposes. Thus, at the touch of a button, the whole wheel can be loosened and taken off the wheelchair easily. With its electric motor and the long-lasting lithium-ion battery, the modern drive concept takes into account the aspect of sustainability.

e-motion

Marc-Gregor Weidt

Designer e-motion M15: Marc-Gregor Weidt

Marc-Gregor Weidt, Jahrgang 1967, schloss nach einer Ausbildung zum Täschner sein Studium als Diplom-Designer im Fachbereich Produktgestaltung an der Fachhochschule für Gestaltung in Schwäbisch Gmünd ab. Schwerpunkte seines Studiums waren unter anderem Investitionsgüterdesign im Bereich der Rehabilitation und die Konzeption und Realisation von Benutzerinterface-Bedienerergonomie. Im Anschluss an sein Studium war er zunächst in den Bereichen visuelle Kommunikation und Produktgestaltung bei design praxis diener tätig. 1995 gründete er mit Concept Design ein Büro für Gestaltung; drei Jahre später firmierte das Unternehmen unter dem Markennamen einmaleins – Büro für Gestaltung um. Im Januar 1999 gründete er die Firma einmaleins – Büro für Projektmanagement, bei der er heute 14 Mitarbeiter beschäftigt. Marc-Gregor Weidt erhielt für seine Arbeiten zahlreiche Auszeichnungen, darunter mehrfach den if Design Award sowie den Designpreis des Landes Baden-Württemberg.

Marc-Gregor Weidt, born in 1967, completed an apprenticeship as a bag maker before graduating with a diploma in design (majoring in product design) from the University of Applied Sciences Schwäbisch Gmünd. The major areas of study during his degree were investment goods in the field of rehabilitation and the conception and implementation of user interface ergonomics. After completing his degree he initially worked in the areas of communication and product design at design praxis diener. In 1995 he founded the design studio Concept Design; three years later he renamed the company to einmaleins – Büro für Gestaltung. In January 1999 he founded the project management company einmaleins – Büro für Projektmanagement, which today has 14 staff. Marc-Gregor Weidt has received several awards for his works, among them the if Design Award, which he received several times, and the Design Prize of the state of Baden-Wurttemberg.

red dot projects Worin liegen Ihrer Meinung nach die besonderen Herausforderungen, die sich heute dem Designer stellen?
Marc-Gregor Weidt Diese liegen im vernetzten Arbeiten. Wir arbeiten als Generalist eng mit Spezialisten zusammen. Das verlangt von uns, sehr tief in den Produktionsprozess einzutauchen und gleichzeitig visionär zu bleiben.

red dot projects Vor welchen speziellen Herausforderungen standen Sie bei der Realisierung Ihres Produktes?
Marc-Gregor Weidt Wie bei fast allen Designaufgaben sind die Herstellungskosten mit dem Pflichtenheft festgeschrieben. Im kreativen Designprozess muss dieser Aspekt von Anfang an bei der Umsetzung mit berücksichtigt werden. Zudem waren eine formal eigenständige, aber gleichzeitig eine dem Rollstuhl angepasste Formgebung Gegensätze, die es zu vereinen galt.

red dot projects Vor dem aktuellen Hintergrund wirtschaftlich schwieriger Zeiten – welche Bedeutung hat gutes Design?
Marc-Gregor Weidt Design differenziert Markenprodukte im Wettbewerb. Bei einer authentischen, emotionalen und doch pragmatischen Gestaltung wirkt sich das positive Empfinden des Nutzers auf den Erfolg dieses Produkts aus.

red dot projects Welche Vorbilder haben Sie?
Marc-Gregor Weidt Was Konsequenz und Lebenseinstellung betrifft, ist dies Otl Aicher und sonst mein Lehrmeister Alfred Koppitz.

red dot projects Warum haben Sie am Wettbewerb REHACARE DESIGN AWARD 2008/2009 teilgenommen und was macht diesen Designwettbewerb besonders?
Marc-Gregor Weidt Der Wettbewerb bietet eine einzigartige und äußerst spannende Möglichkeit der Präsentation im speziellen Umfeld der Reha- und Medizintechnik. Außerdem erreicht man im Zusammenhang mit der REHACARE ein internationales Publikum. Gleichzeitig wollen wir unser besonderes Engagement in Hinsicht auf eine ergonomische, innovative Gestaltung und Konstruktion aufzeigen und dies in einem passenden Umfeld präsentieren.

red dot projects What do you believe are the special challenges designers face today?
Marc-Gregor Weidt They lie in the area of networking. As generalists we are working together closely with specialists. This requires us to immerse ourselves very deeply into the production process and at the same time remain visionary.

red dot projects What special challenges were you faced with during the product realisation process?
Marc-Gregor Weidt As is the case with almost any design task, the manufacturing costs are specified in the technical specification document. This aspect has to be considered in the creative design process right from the start of the implementation. Furthermore, the difficult task was to create a design which was formally independent and at the same time well-suited for the wheelchair.

red dot projects What significance does good design have against the background of the current difficult economic times?
Marc-Gregor Weidt Design differentiates brand name products from the competition. In an authentic, emotional and yet still pragmatic design, the user's positive feelings have an effect on the product's success.

red dot projects Who are your role models?
Marc-Gregor Weidt With regards to consistency and attitude to life it is Otl Aicher, and in other respects my teacher Alfred Koppitz.

red dot projects Why did you take part in the REHACARE DESIGN AWARD 2008/2009 and what makes this competition special?
Marc-Gregor Weidt The competition offers a unique and extremely exciting opportunity to present products in the special area of rehabilitation and medical technology. Furthermore, the REHACARE provides access to an international audience. At the same time we wanted to point out our special dedication with regards to ergonomic, innovative design and construction, and showcase it in a suitable environment.

BEST DESIGN
GEWINNER WINNERS

Herausragende Designleistungen, die in Form und Funktion neue Maßstäbe in ihrer Branche setzen, würdigt die unabhängige Jury mit der Auszeichnung REHACARE DESIGN AWARD: BEST DESIGN.

Outstanding design achievements that set new standards in form and function in their respective industrial sector are distinguished by an independent jury with the REHACARE DESIGN AWARD: BEST DESIGN.

REHACARE
DESIGN AWARD
2008/2009
BEST DESIGN

BEST DESIGN
GEWINNER WINNERS

Produkt Product	Küschall® R33, 2007 Aktiv-Rollstuhl / Active wheelchair
Hersteller Manufacturer	Küschall AG, Witterswil, Schweiz / Switzerland
Werksdesign In-house design	Rainer Küschall, Thomas Kohler
Vertrieb Distributor	Invacare Aquatec GmbH, Isny
Website Hersteller Manufacturer	www.kueschall.com
Website Vertrieb Distributor	www.invacare-aquatec.com

Küschall® R33

Der Starrrahmen-Rollstuhl setzt auf innovative Weise einen Trend in der Branche: Hilfsmittel verstecken sich nicht unauffällig, sondern setzen mit ihrer offensiven Farb- und Formgebung Akzente. Beispielgebend vereint dieser Rollstuhl eine moderne Technologie mit einer auffälligen Gestaltung. Die zentrale technische Komponente ist das neue Stoßdämpfungssystem L.I.S., das die Wirbelsäule entlastet und einen Dämpfungseffekt bei bestmöglicher Energieumsetzung bietet. Es befindet sich unter dem Sitz und fängt diesen auf, um der Nutzerfreundlichkeit, dem Komfort des Fahrers und der Ergonomie Rechnung zu tragen. Das in der Rahmenkonstruktion integrierte Stoßdämpfungssystem vermittelt eine hohe Dynamik und kennzeichnet das Produkt als Rollstuhl für aktive und sportliche Nutzer. Seine kompakten Abmaße lassen ihn auch in beengter Umgebung wendig und agil manövrieren. Auch die Verwendung wertiger Materialien wie Aluminium, Karbon und Titan ist nicht nur Teil der Formensprache, sondern hat funktionstragenden Charakter, indem sie zum geringen Gewicht sowie zur Fahrstabilität beiträgt. Unterschiedliche Varianten für den Rahmen sowie für Sitzkissen und Rückenpolster ermöglichen unter anderem eine individuelle Anpassung des Rollstuhls an die Vorlieben des Nutzers.

The rigid frame wheelchair sets an innovative trend in the industry: aids no longer hide away unobtrusively, but set accents with their offensive colour schemes and designs. This wheelchair combines modern technology with an eye-catching design. Its central technical component is the L.I.S. shock absorption system, which takes weight off the spine and offers an absorption effect with the best possible energy conversion. It is located under the seat and supports it, thus taking into account user-friendliness, comfort for the driver, and ergonomics. The shock absorption system integrated in the frame construction conveys high dynamics and identifies the product as a wheelchair for active and sporty users. Its compact dimensions allow it to manoeuvre with ease and agility, even in tight situations. Carbon and titanium are not only part of the design language, but also play a functional role by contributing to the wheelchair's low weight and driving stability. Different frames, seat and back paddings also allow users to adjust the wheelchair to their individual preferences.

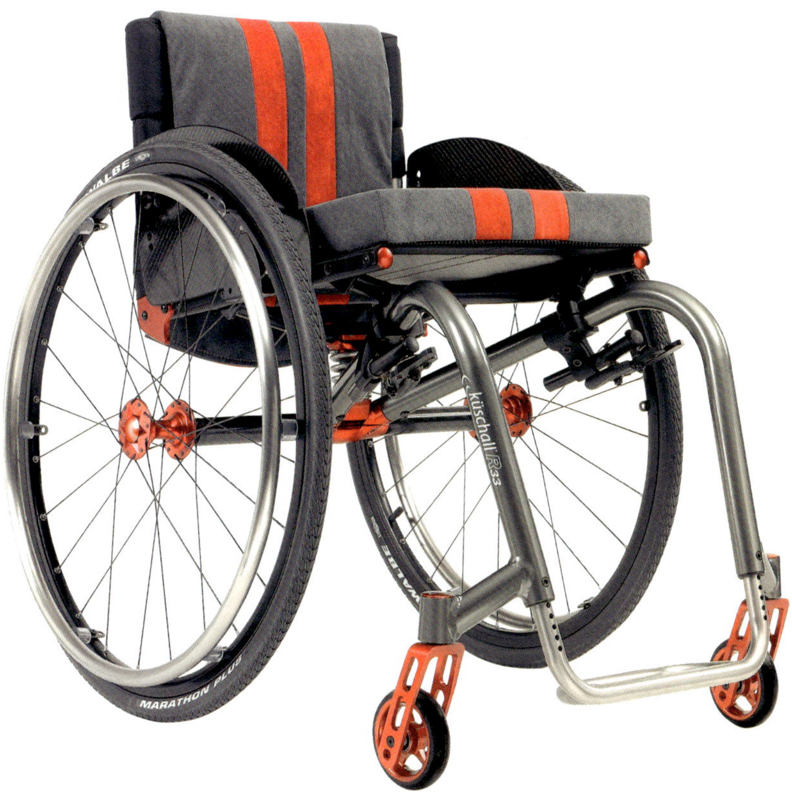
MARATHON PLUS
küschall R33

Rainer Küschall

Designer Küschall R33: Rainer Küschall

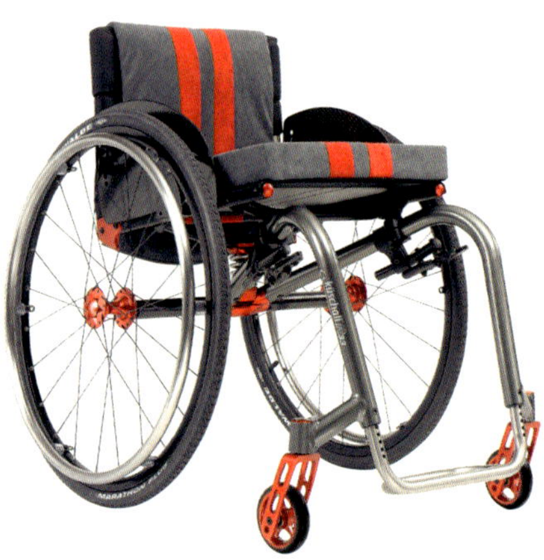

Rainer Küschall, Jahrgang 1947, ist Designer, Rennfahrer und Gründer der Küschall AG mit Sitz in Witterswil/Schweiz. Im Alter von 16 Jahren erlitt er einen Badeunfall, durch den er zum Tetraplegiker wurde. Nachdem er seine spätere Frau kennen gelernt und das Rehacenter verlassen hatte, begann er, eigene Rollstühle zu entwickeln. 1978 gründete er die Firma Küschall of Switzerland und entwickelte den Competition-Rollstuhl, für den er 1986 den Design Award des New Yorker Museum of Modern Art erhielt. In der Folge kamen weitere Rollstuhlserien und Innovationen wie beispielsweise die Vario-Achse, mit der sich der Radsturzwinkel verstellen lässt, auf den Markt, von denen einige ebenfalls mit Preisen ausgezeichnet wurden. Seit 1995 ist die Küschall AG Teil der Invacare-Gruppe, des weltweit größten Anbieters von Hilfsmitteln. Rainer Küschall ist ein erfolgreicher Sportler und nahm mehrfach an den Paralympics teil. Er gewann insgesamt 21 Olympiamedaillen und ist heute ein begeisterter und erfolgreicher Motorsportler.

Rainer Küschall, born in 1947, is a designer, a racing car driver, and the founder of Küschall AG, which is based in Witterswil/Switzerland. At the age of 16 he had a swimming accident which left him a tetraplegic. After meeting his later wife and leaving the rehabilitation centre he started developing his own wheelchairs. In 1978 he founded the company Küschall of Switzerland and developed the Competition Wheelchair, for which he received the New York Museum of Modern Art's Design Award in 1986. In the following years he introduced further successful wheelchair series and innovations to the market such as, for example, the vario axle, which enables drivers to adjust the rear camber angle of the wheels. Some of these series have also won prizes. In 1995 Küschall AG became part of the Invacare group, the world's leading manufacturer and distributor of medical equipment. Rainer Küschall is a successful athlete and has competed in the Paralympics several times. He has won 21 Olympic medals altogether and today is an enthusiastic and successful racing car driver.

red dot projects Worin liegen Ihrer Meinung nach die besonderen Herausforderungen, die sich heute dem Designer stellen?
Rainer Küschall Design muss sich von der Masse abheben, muss Trends setzen. Aber bei all dem muss auch immer die tägliche Handhabung gewährleistet sein. Im Rollstuhlbau ist dies besonders wichtig, weil wir es mit behinderten Menschen zu tun haben.

red dot projects Welche besonderen Aspekte spielen für Sie beim Design für den Rehabilitationsbereich eine Rolle?
Rainer Küschall Funktionale, praktikable, minimale und zweckmäßige Formgebungen kombiniert mit einem modernen, attraktiven und vom Wettbewerb differenzierten Design.

red dot projects Vor dem aktuellen Hintergrund wirtschaftlich schwieriger Zeiten – welche Bedeutung hat gutes Design?
Rainer Küschall Im Prinzip ist die wirtschaftliche Lage für unsere Philosophie nicht ausschlaggebend. Wir möchten unsere Kunden durch innovatives Design in Verbindung mit praktischen technischen Lösungen bestmöglich unterstützen.

red dot projects Was hat Sie zu Ihrem Produkt inspiriert und was ist das Besondere daran?
Rainer Küschall Durch meine Lebens- und Anwendungserfahrung strebe ich ständig nach Innovation und Fortschritt. Das Besondere am Küschall R33 sind die „rahmenlose Geometrie" und die Funktion des L.I.S. Rahmenlose Geometrie bedeutet, dass es so aussieht, als würde der Nutzer über den Rädern schweben. Das L.I.S-System entlastet den Rücken des Nutzers, indem es Schläge und Vibrationen beim Fahren aufnimmt und kompensiert.

red dot projects Welches Wunschprojekt würden Sie gerne einmal realisieren?
Rainer Küschall Rennautos, Segelboote und Flugmaschinen entwickeln.

red dot projects Welche Designphilosophie beeinflusst Ihre Arbeit?
Rainer Küschall Innovative Designs in Verbindung mit praktischen technischen Lösungen.

red dot projects Was inspiriert Sie?
Rainer Küschall Die Intuition allein.

red dot projects What do you believe are the special challenges designers face today?
Rainer Küschall Design has to stand out from the crowd and set trends. However, easy daily use has to be guaranteed. This is particularly important when it comes to wheelchair manufacturing since we are dealing with people with disabilities.

red dot projects What special aspects play a role for you in the field of rehabilitation design?
Rainer Küschall Functional, practical, minimal, and useful forms combined with a modern and attractive design that is distinguished from the competition.

red dot projects Against the background of the current difficult economic times – what significance does good design have?
Rainer Küschall In principle, the economic situation is not crucial for our philosophy. We want to optimally support our customers with innovative design in combination with practical, technical solutions.

red dot projects What inspired you to create this product and what is special about it?
Rainer Küschall Due to my life experience and my personal experiences as a wheelchair user I constantly strive for innovation and progress. What makes the Küschall R33 special is its "frameless geometry" and the function of the L.I.S. Frameless geometry means that it looks as if the user is hovering above the wheels. The L.I.S. system takes the strain off the user's back by absorbing and counteracting shocks and vibrations while driving.

red dot projects What are projects that you would love to work on one day?
Rainer Küschall Racing cars, sailing yachts, and flying machines.

red dot project What design philosophy influences your work?
Rainer Küschall Innovative designs in combination with practical, technical solutions.

red dot projects What inspires you?
Rainer Küschall Intuition alone.

BEST DESIGN
GEWINNER WINNERS

Produkt Product	Krabat jockey, 2008 Therapiestuhl für Kinder / Therapy chair for children
Hersteller Manufacturer	Krabat AS, Hvalstad, Norwegen / Norway
Design	Formel Industridesign AS, Oslo, Norwegen / Norway
Website Hersteller Manufacturer	www.krabat.com
Website Design	www.formel.no

Krabat jockey

Das Lösungskonzept des Therapiestuhls trägt insbesondere der Beweglichkeit des Kindes sowie dem integrativen Aspekt Rechnung. Er zeigt im interaktiven Teil eine ergonomische und organische Form der Gestaltung. Seine funktionstragenden Elemente nehmen sich zugunsten des Nutzers zurück. Sie eröffnen der Betreuungsperson gute Möglichkeiten, den Stuhl an die Bedürfnisse des Kindes anzupassen. Dabei bietet er erhebliche physiologische Vorteile: Der Unterkörper wird durch die breitbeinige Sitzposition und das Fußbrett gestützt, sodass der Oberkörper weniger Unterstützung benötigt. Das Kind gelangt so in eine aktive und dynamische Position, die ihm viele Freiheitsgrade bietet. Passend hierzu verschaffen die stufenlose Verstellung der Sattelneigung sowie die weit einstellbare Höhenregulierung dem Nutzer die Möglichkeit, sowohl in Bodennähe als auch auf einem erhöhten Sitzniveau zu agieren. Aufgrund der zahlreichen Sitzsituationen werden physische Unterschiede zu Nichtbehinderten vermindert. Dank der kompakten Ausführung lässt sich der Therapiestuhl auch bei beengten Platzverhältnissen einsetzen. Zum Transport lässt er sich einfach zusammenlegen und in einer Tasche verstauen.

The concept of this therapy chair takes into account the child's mobility as well as the integrative aspect in particular. The interactive part features an ergonomic and organic design. Its functional elements have an unobtrusive look to benefit the user. They allow caregivers to adjust the chair to the child's needs by, for instance, adjusting the height or the slope of the saddle. The chair offers significant physiological advantages: the lower part of the body is supported by the sitting position with legs wide apart and the footboard, so that the upper body needs less support. Thus users are provided with an active and dynamic position which gives them a high degree of freedom. Furthermore, the continuously adjustable slope of the saddle and the flexibly adjustable height give the user the opportunity to be active close to the floor as well as in an elevated position. Due to the numerous sitting positions physical differences to non-disabled children are reduced. Thanks to its compact design the therapy chair can also be used in confined spaces. It can easily be folded together and put into a bag for transport purposes.

krabat

Aleksander Borgenhov

Designer Krabat jockey: Aleksander Borgenhov

Der 38-jährige Aleksander Borgenhov lebt und arbeitet in der norwegischen Hauptstadt Oslo. Er ist in Norwegen einer der erfolgreichsten Designer des letzten Jahrzehnts. Von 1992 bis 1997 besuchte er die School for Architecture and Design in Oslo, die er als Industriedesigner abschloss. Anschließend war er zwölf Jahre lang als Mitinhaber und Senior-Designberater bei Formel Industridesign AS in Oslo tätig. Aleksander Borgenhov hat zahlreiche Produkte für führende Marken entwickelt, darunter Rehaprodukte, Sportausrüstungsgegenstände, Konsumgüter und Büromöbel. 2009 gründete er mit Powerdesign sein eigenes Unternehmen und übernahm Projekte im Verpackungs-, Büromöbel- und Konsumgüterbereich.

The 38-year-old Aleksander Borgenhov lives and works in the Norwegian capital Oslo. He is one of the most successful designers of the last decade in Norway. From 1992 to 1997 he attended the School for Architecture and Design in Oslo, from where he graduated as an industrial designer. He then worked for Formel Industridesign AS in Oslo for 12 years as co-owner and Senior design consultant. Aleksander Borgenhov has developed several products for leading brands, among them rehabilitation products, sports equipment, consumer goods and office furniture. In 2009 he founded his own company, Powerdesign, and took on projects in the field of packaging, office furniture and consumer goods.

red dot projects Welche besonderen Aspekte spielen für Sie beim Design für den Rehabilitationsbereich eine Rolle?
Aleksander Borgenhov Beim Design denke ich nicht an die Eigenschaft des Produktes als Rehabilitationsprodukt. Ich wähle vielmehr einen Ansatz, den ich auch bei jedem anderen Produkt wähle.

red dot projects Vor dem aktuellen Hintergrund wirtschaftlich schwieriger Zeiten – welche Bedeutung hat gutes Design?
Aleksander Borgenhov Um aus der großen Konkurrenz hervorzustechen, benötigt man innovative Produkte mit einer hohen visuellen Identität. Die Nutzer werden gutes Design immer zu schätzen wissen und einem Unternehmen eine positive Reputation geben, die für Geld nicht zu kaufen ist.

red dot projects Worin liegen Ihrer Meinung nach die besonderen Herausforderungen, die sich heute dem Designer stellen?
Aleksander Borgenhov Die Herausforderung liegt darin, umweltfreundliche und nachhaltige Produkte zu gestalten, die eine Mehrheit der Konsumenten ansprechen.

red dot projects Was hat Sie zu Ihrem Produkt inspiriert und was ist das Besondere daran?
Aleksander Borgenhov Die Zusammenarbeit mit dem Krabat-Team war sehr inspirierend. Sie sind bestrebt, die Welt der Rehaprodukte zu verändern. Krabat hatte schon das ursprüngliche Produktkonzept entwickelt, als wir zum Team stießen und das Industrial Design beisteuerten. Wir wollten das sitzende Kind in den Mittelpunkt stellen und nicht das Produkt.

red dot projects Vor welchen speziellen Herausforderungen standen Sie bei der Realisierung Ihres Produktes?
Aleksander Borgenhov In diesem Fall galt es zahlreiche technische Herausforderungen und Funktionseigenschaften bei begrenzten Möglichkeiten in der Produktion zu bewältigen.

red dot projects What are the special aspects that play a role for you in the field of rehabilitation?
Aleksander Borgenhov When I create a design I don't think of the product specifically as a rehabilitation product. Instead I choose the approach I use for any other product.

red dot projects Against the backdrop of current difficult economic times – how important is good design?
Aleksander Borgenhov In order to stand out from the competition you need innovative products with a strong visual identity. Users will always appreciate good design, thus companies get a good reputation that money cannot buy.

red dot projects What do you believe are the special challenges designers face today?
Aleksander Borgenhov The challenge is to design environmentally friendly and sustainable products which appeal to a majority of consumers.

red dot projects What inspired you to create this product and what makes it special?
Aleksander Borgenhov Collaborating with the Krabat team was very inspiring. They are striving to change the world of rehabilitation products. Krabat had already developed the original product concept when we joined the team and contributed the industrial design. Our aim was to make the sitting child, and not the product, the primary focus.

red dot projects What were the special challenges you were faced with during the realisation process of your product?
Aleksander Borgenhov In this case we had to find solutions for several technical challenges and functional features while being faced with limited production possibilities.

BEST DESIGN
GEWINNER WINNERS

Produkt Product	Conveen Optima, 2007 Kondom-Urinal (externer Katheter) / Urisheath
Hersteller Manufacturer	Coloplast A/S, Humlebaek, Dänemark / Denmark
Werksdesign In-house design	Henrik Lindenskov Nielsen
Website Hersteller Manufacturer	www.coloplast.dk

Conveen Optima

Körperliche Leiden, die sich auf den Intimbereich beziehen, erfordern eine besondere Sensibilität hinsichtlich ihrer Gestaltung. Das Kondom-Urinal zeigt eine Lösung, die diesen Aspekt in einer besonderen Weise berücksichtigt und eine gewisse Selbstverständlichkeit in den Alltag bringt. Zunächst verdeutlicht die markante Formgebung des Urinals unmissverständlich seine Funktion, während die Weichheit des Materials die angenehme Handhabung sowohl für Selbstanwender wie auch Pflegepersonal betont. Die überlegt ausgearbeiteten Details des Produktes tragen dazu bei, dass die Anwendung so einfach wie möglich gehalten wird, wodurch Stresssituationen vermieden werden. Hier ist der doppelte Abrollstreifen zu nennen, der das gleichmäßige hygienische Abrollen ohne Falten gewährleistet und die Berührungspunkte verringert. Weiterhin greift das Wellenprofil am Abflussstutzen nicht nur die Gestaltung des Ziehharmonika-Knickschutzes auf, sondern ermöglicht die sichere Befestigung des Beutels. Um den Tragekomfort des Kondom-Urinals zu erhöhen, wurde ein Kleber verwendet, der sowohl eine sichere Haftung zulässt als auch hautfreundlich ist. Abgerundet wird das Gestaltungskonzept des Produktes durch seine Verpackung. Ihre grüne Farblichkeit verleiht dem Hilfsmittel eine natürliche Ausstrahlung, die etwaige Unsicherheiten beim Benutzer abzubauen hilft. Sie zeigt sich diskret und funktional, da sie sich unauffällig mitführen und mit einer Hand öffnen lässt.

Physical conditions that affect the genital area require special sensitivity with regards to design. The condom urinal presents a solution which takes into account this aspect in a special way, bringing a certain degree of naturalness to everyday life. The distinctive form of the urinal clearly indicates its function, while the softness of the material emphasises the pleasant use to individuals and nursing staff alike. The product's sophisticated details contribute to making its use as easy as possible, thus preventing stressful situations. In this context the double unrolling strip is worth mentioning, which allows consistent unrolling without creasing, while simultaneously reducing the area of contact. Furthermore, the discharge nozzle's wave profile not only continues the concertina creasing protection, but also provides safe attachment of the bag. In order to increase the user's comfort when wearing the condom urinal, a glue has been used which provides safe grip while being skin-friendly at the same time. The product's design concept is completed through its packaging. Its green colour gives the aid a natural look, which helps to reduce any possible feelings of unease in the user. It is discreet and functional, since it can be taken along without attracting attention and can be opened with one hand.

30
Conveen
Coloplast

Henrik Lindenskov Nielsen

Designer Conveen Optima: Henrik Lindenskov Nielsen

Henrik Lindenskov Nielsen, 48, lebt in der Nähe von Kopenhagen. Nach seiner Ausbildung zum chemischen Ingenieur war er ab 1984 in der Forschung und Entwicklung für Lichtleiter und Leiter zur optischen Nachrichtenübertragung tätig und anschließend bei Boston Scientific für die Forschung und Entwicklung von Ballonkathetern, Stents und für künstliche Blutgefäße zuständig. Beim Unternehmen Coloplast beschäftigte er sich in der Folge mit der Forschung und Entwicklung im Bereich der Inkontinenzprodukte. Heute betreut er federführend die weltweite Forschungsabteilung des Unternehmens. Sein besonderes Interesse gilt der kreativen Nutzung von neuen und bestehenden Technologien, um Kundenbedürfnissen gerecht zu werden.

Henrik Lindenskov Nielsen, aged 48, lives near Copenhagen. After training as a chemical engineer he worked on the research and development of optical fibres and optical information cables from 1984 and then at Boston Scientific on the research and development of ballon catheters, stents and for artificial blood vessels. At Coloplast, he was then involved in research and development in the area of incontinence products. Today he is in charge of the global research and development department. His special interest is in the creative use of new and existing technologies in order to satisfy customers' needs.

red dot projects Vor welchen speziellen Herausforderungen standen Sie bei der Realisierung Ihres Produktes?
Henrik Lindenskov Nielsen Die größte Herausforderung bei der Entwicklung war, dass wir mit neuen Materialien und Produktionsprozessen arbeiten mussten und gleichzeitig das Zusammenwirken der Materialien so gestalten mussten wie beabsichtigt. Außerdem war es eine Herausforderung, die Einmalverpackung diskret und umweltfreundlich zu gestalten und gleichzeitig dem Kundenbedürfnis gerecht zu werden, dass auch Anwender mit eingeschränkter Handfunktion das Produkt einfach anwenden können.

red dot projects Vor dem aktuellen Hintergrund wirtschaftlich schwieriger Zeiten – welche Bedeutung hat gutes Design?
Henrik Lindenskov Nielsen Ein gutes Design ist sehr wichtig, darum legen wir auch sehr viel Wert darauf. Ein innovatives Design wird in Zukunft ein „Muss" sein. Wenn wir auf diesem Gebiet die Nase vorne haben, haben wir gute Chancen, auf dem Markt zu bestehen.

red dot projects Worin liegen Ihrer Meinung nach die besonderen Herausforderungen, die sich heute dem Designer stellen?
Henrik Lindenskov Nielsen Ein guter Designer muss in der Mitte des Produktentwicklungsprozesses stehen. Er muss die Situation der Anwender sehr gut verstehen, was einen intensiven Dialog mit den Kunden erfordert. Auf der anderen Seite muss er aber auch im intensiven Dialog zu den Spezialisten stehen, die innovative Lösungen für Produkteigenschaften, Materialien und den Produktionsprozess liefern.

red dot projects Was inspiriert Sie?
Henrik Lindenskov Nielsen Ich werde sowohl von sehr technischen Dingen wie auch von der Vielfältigkeit der Natur inspiriert. Mit Menschen zusammenzuarbeiten, die unterschiedlichste Fähigkeiten haben, ist ebenfalls eine meiner Inspirationsquellen. Auch meine Hobbys, Gitarre spielen, Joggen, Skifahren oder Snowboarden, geben mir neue Energie für meine Arbeit.

red dot projects What special challenges were you faced with in the realisation process of your product?
Henrik Lindenskov Nielsen The biggest challenge of the development was that we had to work with new materials and manufacturing processes while at the same time we had to make the interaction between the material work as intended. Furthermore, it was a challenge to create a discreet and environmentally friendly design for the single use packaging while at the same time meeting the customers' requirement that also users with limited hand function could use the product easily.

red dot projects What significance does good design have against the background of the current difficult economic times?
Henrik Lindenskov Nielsen Good design is very important; that is why we place a high value on it. Innovative design will be a must in future. If we have the edge on the competition in this area, then we have a good chance of surviving in the market.

red dot projects What do you believe are the special challenges designers face today?
Henrik Lindenskov Nielsen A good designer has to be at the centre of the product development process. He has to understand the users' situation very well, which requires an intensive dialogue with the customer. On the other hand, he also has to engage in an intensive dialogue with the specialists who provide innovative solutions for product qualities, materials, and the manufacturing process.

red dot projects What inspires you?
Henrik Lindenskov Nielsen I am inspired by very technical things as well as nature's diversity. Working together with people who have different skills is also one of my sources of inspiration. My hobbies, such as playing the guitar, jogging, skiing, and snowboarding, also give me new energy for my work.

GOOD DESIGN
GEWINNER WINNERS

Die Auszeichnung REHACARE DESIGN AWARD: GOOD DESIGN wird für besondere Designleistungen in Form und Funktion vergeben.

The distinction REHACARE DESIGN AWARD: GOOD DESIGN is awarded for special design achievements in form and function.

REHACARE
DESIGN AWARD
2008/2009
GOOD DESIGN

GOOD DESIGN
GEWINNER WINNERS

Produkt Product	Invacare® Comet, 2008 Scooter
Hersteller Manufacturer	Invacare Aquatec GmbH, Isny
Werksdesign In-house design	Diana Nerving, Norbert Mesch
Website Hersteller Manufacturer	www.invacare-aquatec.com

Invacare® Comet

Ein aktiver und selbstbestimmter Lebensstil findet seinen Ausdruck in einer hohen Mobilität. Elektromobile bieten im Alltag die Freiheit der unabhängigen Fortbewegung und erhöhen dadurch die Selbstständigkeit ihrer Nutzer. Der Scooter zeigt sich hier als herausragendes Beispiel für eine gelungene Synthese von Funktionalität und Formgebung. Seine formale Eleganz verbindet sich mit einer intelligenten Form der Benutzerführung, die aufgrund der Verwendung symbolischer Elemente auch international verständlich ist. Ein wesentlicher funktionaler Aspekt ist die Interaktion des Anwenders mit dem Produkt, die hier so einfach wie möglich gehalten und auf das Notwendigste reduziert wird. So verringert sich beispielsweise die Geschwindigkeit in Kurvenfahrten automatisch, was den Benutzereingriff bei der Fahrt überflüssig macht. Das klar strukturierte Bedienfeld mit den erhabenen Tasten und der deutlich lesbaren Ladestandsanzeige trägt den Ansprüchen an Ergonomie und Gestaltung Rechnung. Durch die Einzelradfederung sowie die großen Räder überwindet das Elektromobil zuverlässig Bodenunebenheiten, Hindernisse oder Bordsteine.

High mobility is an essential component of an active and independent lifestyle. Electronic scooters offer independent mobility in daily life, thus increasing the user's autonomy. This scooter presents itself as an excellent example of a successful synthesis of functionality and form. Its formal elegance is linked with an intelligent user guidance, which can be understood internationally thanks to its use of symbolic elements. One essential functional aspect is the interaction between the user and the product, which has been designed to be as easy as possible and has been reduced to the essentials. The speed, for instance, is automatically reduced around corners, which makes any user intervention while driving unnecessary. The well-structured control panel with its raised buttons and the clear state-of-charge display take into account ergonomics and design. Thanks to the wheels' independent suspension as well as their large size, uneven surfaces, obstacles and kerbstones can be negotiated safely.

Produkt Product	Viper Plus, 2007 Elektro-Rollstuhl / Electric wheelchair
Hersteller Manufacturer	Days Healthcare GmbH & Co. KG, Bad Oeynhausen
Werksdesign In-house design	Achim Kaesberg
Website Hersteller Manufacturer	www.dayshealthcare.de

Viper Plus

Das auffälligste Element des Elektro-Rollstuhls ist das hintere Rahmenelement, das gleichzeitig als Schutz für die Batterieschublade dient. Es ist nicht nur funktionstragend, sondern verleiht dem Produkt eine sportlich-robuste Anmutung. Auch die aufgrund ihrer abgesetzten Farbgebung sehr präsenten Profilreifen unterstreichen die Gesamtanmutung des Produkts. Der Rahmen ist in Schwarz oder Orange erhältlich, sodass er wahlweise eine auffällige oder zurückgenommene Optik bietet. Die geradlinig ausgeführte Sitzpolsterung und die Armlehnenpolster aus Integralschaum wirken einladend und bequem. Der Gebrauchswert des Rollstuhls resultiert aus den intelligenten Details, wozu die wegschwenkbaren Blinker, die Beleuchtung und das ergonomisch ausgeformte Bediengerät zählen. Zudem lässt sich das Produkt über vielfältige Optionen an die unterschiedlichen Bedürfnisse seiner Nutzer anpassen: Zum einen bietet es die Möglichkeit, die Sitzbreite ebenso wie die Arm- und Beinlehnen variabel einzustellen, zum anderen kann der Nutzer die Sitzneigung individuell wählen, um neben einem hohen Maß an Komfort eine bestmögliche Körperkontrolle zu erzielen.

The most eye-catching element of this electric wheelchair is its rear frame element which also serves as a protection for the battery drawer. It is not only a functional element, but gives the product a sporty and robust appearance. Thanks to their contrasting colour, the wheels are very noticeable, additionally emphasising the product's overall appearance. The frame is available in black or orange, offering the choice between an unobtrusive and an eye-catching look. The straight seat padding and the armrests made of integral foam appear inviting and comfortable. The wheelchair's utility is a result of the intelligent details, which include swivelling indicators, lights, and an ergonomically shaped control element. Furthermore, the product can be adjusted to the user's individual requirements thanks to many options: on the one hand, the seat width as well as the armrests and legrests can be adjusted, on the other hand the user can choose the angle of the seat in order to achieve the best possible body control in addition to a high degree of comfort.

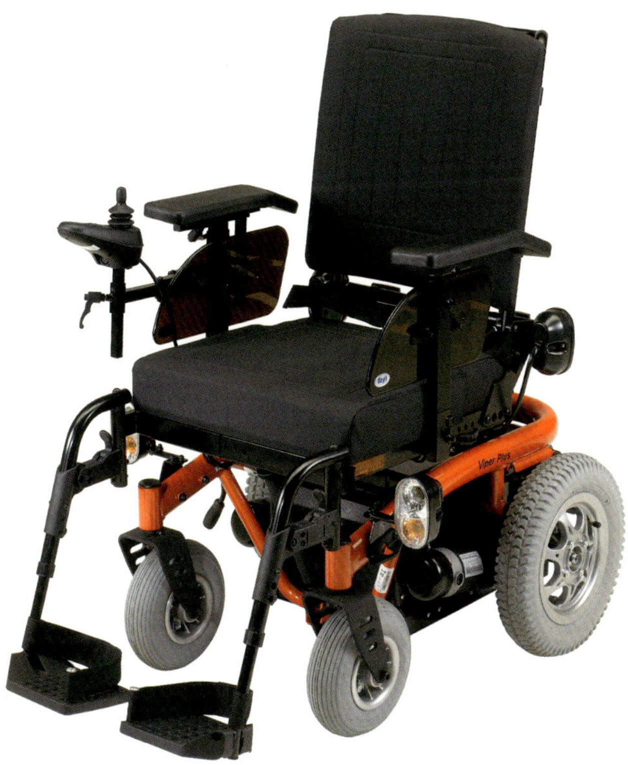

GOOD DESIGN
GEWINNER WINNERS

Produkt Product: Xeno, 2007
Elektro-Rollstuhl / Power wheelchair
Hersteller Manufacturer: Otto Bock Mobility Solutions GmbH, Königsee
Werksdesign In-house design: Oliver Kretschmer, Heinrich Perk
Website Hersteller Manufacturer: www.ottobock.de

Xeno

Menschen, die in ihrer Bewegung eingeschränkt sind, benötigen Hilfsmittel, die neben der für sie geeigneten Funktionalität eine hohe Sicherheit im Gebrauch vermitteln. Der Elektro-Rollstuhl mit Aufstehhilfe erreicht dies dank seiner in sich schlüssigen Gestaltung. Diese lässt die Funktionalität klar in den Vordergrund treten und schafft so für seinen Benutzer eine angenehme Verlässlichkeit. Eine offensichtliche Solidität und die konsequent ausgearbeiteten Details tragen dazu bei, dass die Möglichkeiten, die der Rollstuhl bietet, leicht zu erkennen und zu nutzen sind. So kann der Nutzer die Sitzbreite und -tiefe individuell verstellen oder die Rückenlehne und Sitzkantelung mit dem Joystick variieren. Durch die Steuerung mit der Einzelradzwangslenkung eignet sich der Elektro-Rollstuhl auch für beengte Platzverhältnisse. Im Alltag erweitert die Aufstehhilfe den Aktionsradius des Nutzers, indem er die Bewegung in der Vertikalen unterstützt. Dies ermöglicht dem Anwender eine neue Teilhabe an seiner Umwelt – ein Aspekt, der positive Auswirkungen sowohl auf das körperliche wie auch das psychische Wohlbefinden hat.

People with movement restrictions need aids that provide a very safe use in addition to individually required functionality. The power wheelchair with stand-up function achieves this through its coherent design. It allows its functionality to come to the fore, thus providing the user with a pleasant reliability. The obvious robustness and the sophisticated details contribute to users easily recognising and using the wheelchair's functions. They can individually adjust the seat depth and width and modify the backrest and seat tilt using a joystick. Due to its single-wheel forced steering, the electric wheelchair is also suited for use in tight spaces. In daily situations the stand-up function increases the user's range of movement by supporting vertical manoeuvres. This enables users to participate in their environment in a new way – an aspect with positive effects on physical as well as psychological well-being.

Produkt Product	BRAVOracer, 2007 Manueller Rollstuhl / Manual wheelchair
Hersteller Manufacturer	Otto Bock Mobility Solutions GmbH, Königsee
Werksdesign In-house design	Oliver Kretschmer, Alexander Müller
Website Hersteller Manufacturer	www.ottobock.de

BRAVOracer

Der Starrrahmen-Rollstuhl stellt dank seiner grafischen und farblichen Gestaltung weit mehr als ein Hilfsmittel für behinderte Kinder dar. Formale Elemente, die von modischen Produkten stammen, werden in das Hilfsmittel integriert, sodass eine Verbindung zwischen beiden Welten geschaffen wird. Das Gerät strahlt eine positive, agile Lebensfreude aus und ermöglicht dem jungen Anwender ein Maximum an Stabilität und Anpassbarkeit. Der geschwungene Vorderrahmen mit den breiten Auslegern verbindet lebhafte Dynamik mit therapeutisch sinnvollen Elementen und sichtbarer Robustheit. Die zahlreichen Einstellmöglichkeiten sind harmonisch integriert, sodass ein spielerisches Gesamtbild geschaffen wird. Das Gerät wurde anhand kindlicher Körper- und Wachstumsdaten entwickelt, um sich den Bedürfnissen von Kindern optimal anzupassen. Anforderungen der Therapeuten finden sich ebenso berücksichtigt wie die Wünsche der Eltern und Begleitpersonen. Details wie der im Seitenteil integrierte Bremshebel oder das breite Fußbrett belegen die kindgerechte Funktionalität des Rollstuhls. Durch optionale Rahmenfarben und verschiedene Ausstattungselemente kann der Nutzer dem Produkt selbst eine persönliche Note verleihen.

Thanks to its graphic design and colour scheme, the rigid frame wheelchair is much more than just an aid for disabled children. Formal elements, which come from fashionable products, have been integrated in the aid creating a connection between both worlds. The device conveys a positive, agile zest for life and provides the young user with a maximum of stability and flexibility. The curved frontal frame with its wide arms combines lively dynamics with therapeutically useful elements and visible robustness. The numerous adjustment options have been harmoniously integrated, creating a playful overall appearance. The device was developed on the basis of children's body and growth data to adapt to children's needs in the best possible way. Requirements of the therapists have been taken into account as well as the wishes of parents and accompanying persons. Details such as the brake lever integrated in the side frame and the wide footboard demonstrate that the functionality of the wheelchair is particularly suitable for children. Optional frame colours and features allow users to personalise their product.

GOOD DESIGN
GEWINNER WINNERS

Produkt Product	Martin Conquest 1150 R, 2007 High-Performance-Motorrad / High-performance motorcycle
Hersteller Manufacturer	Funwheelz, Haaren, Niederlande / The Netherlands
Werksdesign In-house design	Martin Allen
Website Hersteller Manufacturer	www.funwheelz.nl

Martin Conquest 1150 R

Dieses leistungsfähige Motorrad erlaubt Behinderten in einen Bereich vorzudringen, der bislang Menschen ohne körperliche Einschränkungen vorbehalten war. Setzte das Motorradfahren bisher eine hohe Fahrtauglichkeit und körperliche Unversehrtheit voraus, so bietet dieses auf Rollstuhlfahrer zugeschnittene Lösungskonzept neue Möglichkeiten der Fortbewegung. Dies wird durch die kraftvolle, dynamisch anmutende Gestaltung des Trikes deutlich: Sie vermittelt ein hohes Maß an Eigenständigkeit und Lebensfreude, die in eine angemessene Formensprache übersetzt wurden. Rollstuhlfahrer können über eine automatisierte Rampe in das Motorrad rollen und anschließend die rückwärtige Verriegelung über eine Drucktaste selbst schließen. Der Rollstuhl verschwindet hinter der ausladenden Verkleidung, sodass die Behinderung des Fahrers in den Hintergrund tritt. Das Trike besitzt sechs Gänge sowie einen Rückwärtsgang. Hydraulisch geregelte Scheibenbremsen vorne und hinten sind über einen Hebel zu bedienen.

This high-performance motorcycle gives disabled people access to an area previously reserved for people without physical restrictions. While in the past riding a motorcycle required a high level of fitness and full physical ability, this new concept developed for wheelchair users offers new opportunities in terms of mobility. This is clearly conveyed by the trike's powerful, dynamic design: it conveys a high degree of independence and a zest for life, which has been translated into an appropriate design language. Wheelchair users can roll into the motorcycle via an automatic ramp and then close the back lock themselves by pushing a button. The wheelchair disappears behind the wide fairings so that the rider's disability fades into the background. The trike has six forward gears and one reverse gear. The front and rear hydraulic disc brakes are lever-operated.

Produkt Product	Monza Reha, 2008 Autokindersicherheitssitz / Children's safety seat for cars
Hersteller Manufacturer	Thomas Hilfen für Körperbehinderte GmbH & Co. Medico KG, Bremervörde
Werksdesign In-house design	Werner Schönbeck
Design	RECARO Competence Center Child Safety Jürgen Schrödel, Marktleugast
Website Hersteller Manufacturer	www.thomashilfen.de
Website Design	www.recaro.de

Monza Reha

Der Autositz für Kinder mit Handicaps ist ein Beispiel für die Weiterentwicklung eines bestehenden Produktkonzeptes für eine besondere Zielgruppe und wirkt damit als ein hochtechnisches integratives Produkt, das nicht vordergründig als Hilfsmittel für Behinderte erscheint. Er zeigt funktional ergonomische Lösungsansätze, die durch unterschiedliche technische Details verbunden sind. Als Basis dient ein besonders sicherer Recaro-Autositz, in den Additive wie verschiedene Thorax- und Kopfpelotten, eine Fußstütze, ein Abduktionsblock und ein aufsteckbarer Tisch integriert werden können. Die zusätzlichen Komponenten fügen sich harmonisch in die Gestaltung ein. Die Ausführung der Polster betont den ergonomischen Anspruch und signalisiert einen hohen Komfort. Polster, Styroporkörper und Außenschale sind so gestaltet, dass durch ein spezielles Belüftungssystem das Schwitzen des Kindes verhindert wird. Die Kopfstütze ist mittels integriertem Luftkissen stufenlos in der Tiefe zu verstellen. Dadurch bleibt der Kopf des schlafenden Kindes sicher in einer etwas nach hinten geneigten Position.

The car seat for children with disabilities is an example of the development of an existing product concept for a special target group and thus becomes a high-tech integrative product, which on the surface does not look like an aid for people with disabilities. It features functionally ergonomic solutions, which are linked through different technical details. The basis is a particularly safe Recaro car seat with integrated thorax and head pads, a footrest, an abduction block, and a mountable table. The additional components all blend in harmoniously with the design. The execution of the upholstery emphasises ergonomic aspects and signals a high degree of comfort. The upholstery, the polystyrene body, and the outer shell feature a special venting system which prevents the child from sweating. Thanks to its integrated air cushion, the headrest's depth can be continuously adjusted. Thus the head of the sleeping child remains in a safe position slightly tilted backwards.

GOOD DESIGN
GEWINNER WINNERS

Produkt Product	Veigel Commander Classic, 2008 Integrierte Fernbedienung für behinderte Autofahrer Integrated remote control for disabled motorists
Hersteller Manufacturer	Veigel GmbH + Co. KG, Künzelsau
Design	Thamm Design, Langenburg
Website Hersteller Manufacturer	www.veigel-automotive.de
Website Design	www.thamm-design.de

Veigel Commander Classic

Ein Hauptaugenmerk bei der Entwicklung von Reha-Hilfsmitteln liegt auf einer Erhöhung der Mobilität behinderter Menschen. Ein Höchstmaß an selbstbestimmter Fortbewegung und Integration in den Alltag wird erreicht, wenn Menschen mit Handicap das Autofahren ermöglicht wird. Die integrierte Fernbedienung für Autofahrer wurde im Hinblick auf Ergonomie, Komfort und Sicherheit konzipiert. Sie zeigt eine zurückhaltende Formensprache und fügt sich dadurch in das Interieur moderner Fahrzeuge harmonisch ein. Das Gerät ist auf eine sehr funktionale und klare Gestaltungssprache reduziert. Seine abgerundet ausgeformten Oberflächen vermitteln das Gefühl eines Handschmeichlers, der dennoch eine seriöse Funktion erfüllt. Aufgrund ihrer Modularität eignet sich die Fernbedienung für den Einbau in zahlreiche Fahrzeuge. Das Handgerät verfügt neben den Funktionen Gas und Bremse über eine integrierte Fernbedienung für die Funktionen Blinker, Warnblinker, Licht, Fernlicht, Lichthupe, Scheibenwischer, Hupe und Scheinwerfer. Alle Funktionen des Gerätes lassen sich mit einem Finger bedienen, wobei der Anwender die Stellung der Hand auf dem Handgriff beibehalten kann.

Increasing the mobility of disabled people is a major focus in the development of rehabilitation aids. The highest possible degree of independent mobility and integration into daily life is achieved when people with disabilities are able to drive cars. The integrated remote control for motorists was developed with ergonomics, comfort and safety in mind. It features an unobtrusive language of forms, thus harmoniously blending in with the interior of modern vehicles. The device has been reduced to a very functional and clear design language. Its rounded surfaces convey the sensation of a smooth stone to the user while providing an important function. Along with the accelerator and brake functions, the hand-control features an integrated remote control to operate the indicators, hazard lights, light, drive light, headlight flasher, wipers, horn, and headlights. All functions can be operated with just one finger, allowing the user to keep one hand resting on the handle.

Produkt Product: Veigel Commander Compact, 2008
Integrierte Fernbedienung für behinderte Autofahrer
Integrated remote control for disabled motorists
Hersteller Manufacturer: Veigel GmbH + Co. KG, Künzelsau
Design: Thamm Design, Langenburg
Website Hersteller Manufacturer: www.veigel-automotive.de
Website Design: www.thamm-design.de

Veigel Commander Compact

Die integrierte Fernbedienung für behinderte Autofahrer erfüllt ein wichtiges Kriterium für moderne Hilfsmittel: Sie erweitert die Anwenderschnittstellen von Produkten für Nichtbehinderte derart, dass eine Nutzung von behinderten Menschen möglich ist. Dies erreicht sie über eine optische und funktionale Kompatibilität. Dank ihrer zurückhaltenden formalen Gestaltung fügt sie sich harmonisch in das Interieur aller modernen Fahrzeuge ein. Aufgrund ihrer reduzierten Gestaltung treten ergonomische und sicherheitsrelevante Grundsätze in den Vordergrund. Zum einen ist der Griff so geformt, dass der Fahrer seine Hand während der Fahrt darauf ablegen kann. Zum anderen zeigt das Produkt abgerundete Linien, sodass die Verletzungsgefahr im Falle eines Unfalls reduziert wird. Das Handgerät verfügt neben den Funktionen Gas und Bremse über eine integrierte Fernbedienung für die Funktionen Blinker, Warnblinker, Licht, Fernlicht, Scheibenwischer, Lichthupe und Hupe. Die Modularität des Produktes verleiht ihm einen zusätzlichen Wert, da es sich für den Einbau in zahlreiche Fahrzeuge eignet.

The integrated remote control for disabled motorists fulfils an important criterion for modern aids: it increases the user interfaces of products for non-disabled people in such a way that it enables disabled people to use them. This is achieved by visual and functional compatibility. Thanks to its unobtrusive language of form, the integrated remote control for disabled motorists blends in harmoniously with the interior of modern vehicles. Due to the streamlined design, aspects of ergonomics and safety-relevance come to the fore. On the one hand, the handle's specific shape allows the driver to rest his hand on it while driving. On the other hand, the product features rounded lines reducing the injury risk in case of an accident. Along with the accelerator and brake functions, the hand-control features an integrated remote control to operate indicators, hazard lights, light, drive light, wipers, headlight flasher, and horn. The product's modularity gives it additional value since it makes it suitable for retrofitting to many vehicles.

GOOD DESIGN
GEWINNER WINNERS

Produkt Product	Thevo Therapy, 2008 Therapiestuhl für Kinder / Therapy chair for children
Hersteller Manufacturer	Thomas Hilfen für Körperbehinderte GmbH & Co. Medico KG, Bremervörde
Werksdesign In-house design	Julia Zacher (Produktdesignerin), Monika Zupke, Michael Tiedemann
Website Hersteller Manufacturer	www.thomashilfen.de

Thevo Therapy

Bewegung ist ein Grundprinzip des sich körperlich, geistig und seelisch entwickelnden Lebens. Kinder begreifen sich und ihre Umwelt über Bewegung, was gerade auch für behinderte Kinder in besonderer Weise gilt. Das Konzept dieses Therapiestuhls trägt dem Bewegungsdrang von Kindern Rechnung. Er unterstützt die aktive Bewegung, wenn sich das Kind in einer Sitzposition befindet. Hierzu tragen die stabilisierenden und stimulierenden Flügelfedern im Rückenbereich bei, die auf kleinste Eigenbewegungen des Kindes reagieren. Diese minimalen Impulse, ausgesandt von den beweglichen Auflagepunkten der Flügelfedern, aktivieren die Kinder und motivieren sie zu mehr Bewegung. Aufgrund seiner offenkundigen Funktionalität findet der Therapiestuhl eine Formensprache, die den Therapeuten und Betreuungspersonen die Möglichkeit gibt, ihn in seinem ganzen Umfang anzuwenden. Für die jungen Benutzer wird über die kindgerechte Farbgebung des Bezuges eine positive Vertrautheit geschaffen. Sie lenkt von der Eigenschaft des Produktes als Reha-Hilfsmittel ab und fördert die Akzeptanz seitens seiner Benutzer. Die beweglichen Elemente im Rücken lassen sich ebenso wie Armlehnen, Höhe und Sitzneigung individuell an den jeweiligen Nutzer anpassen.

Movement is a fundamental principle of physical, mental and emotional human development. Children understand themselves and their environment through movement; this applies to disabled children in particular. The concept of this therapy chair takes into account the children's need to move. It supports active movement when the child is in a sitting position. The stabilising and stimulating wing-shaped suspension elements in the back area contribute to this by responding to the smallest movements of the child. These minimal impulses sent off by the wing suspension element's flexible support points activate the children and motivate them to move more. Due to its obvious functionality the therapy chair features a design language which gives therapists and nursing staff the opportunity to use all of its functions. The cover's child-oriented colour scheme creates a positive familiarity for the young users. It detracts attention from the product's function as a rehabilitation aid and increases acceptance by its users. The movable elements in the back as well as the armrests, the height, and the seat slope can be adjusted to the user's individual needs.

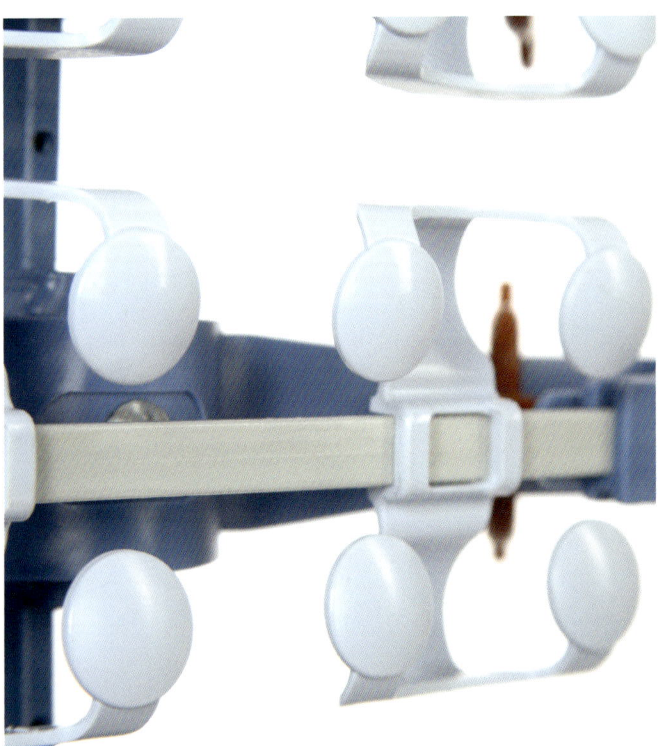

Produkt Product	Krabat pilot, 2007 Krabbelhilfe / Crawling aid
Hersteller Manufacturer	Krabat AS, Hvalstad, Norwegen / Norway
Design	Formel Industridesign AS Aleksander Borgenhov, Oslo, Norwegen / Norway
Website Hersteller Manufacturer	www.krabat.com
Website Design	www.formel.no

Krabat pilot

Kinder besitzen einen hohen Bewegungsdrang, sodass bei ihnen der Wunsch nach Mobilität besonders ausgeprägt ist. Diesem Aspekt trägt das Hilfsmittel Rechnung, indem es behinderten Kindern in den Vierfüßlerstand hilft und damit die Voraussetzungen für das selbstständige Krabbeln schafft. Durch seine sorgfältig ausgearbeiteten Details erreicht das Produkt eine hohe Benutzerfreundlichkeit, sodass es sich sicher und einfach handhaben lässt. Beispielhaft steht hier der Plastikmantel, der die Räder schützt, das Gerät stabilisiert und verhindert, dass das Kind seine Finger bei der Fortbewegung einklemmt. Seine rundliche Formensprache in Verbindung mit der fröhlichen Farbgestaltung signalisiert Bewegung, Geschwindigkeit und positive Gefühle. Es motiviert damit neben dem Kind selbst auch die Betreuenden und die Eltern. Die Krabbelhilfe besitzt ein integriertes Federsystem, das die Hebung des Beckens bewirkt und das Beugen der Hüften ermöglicht. Ihre dynamische Hebebewegung variiert mit der Bewegung des Kindes, sodass es seine Aktionen aktiv unterstützt. Vier verschiedene Höhen und mehrere unterschiedliche Härtegrade des Federsystems stehen zur Verfügung, um die Krabbelhilfe an die Bedürfnisse des Kindes anzupassen. Dank der verschiedenen Farben und Stoffmuster lässt sich das Produkt ebenfalls an den Geschmack der Nutzer anpassen.

Children have a great urge to move; therefore their desire for mobility is particularly strong. The aid takes this aspect into account by helping disabled children into the all-fours position, thus providing the prerequisites for independent crawling. Thanks to its sophisticated details, the product is very user-friendly so that it can be used safely and easily. The plastic coating, for instance, protects the wheels, stabilises the device and prevents the child from getting its fingers caught in the wheels while moving. Its round design language in combination with its happy colour scheme signals movement, speed, and positive emotions. Thus, along with the child, it also motivates caregivers and parents. The crawling aid has an integrated suspension system which lifts the pelvis and facilitates the bending of the hip. Its dynamic lifting movements adapt to the child's movements, thus actively supporting its activities. Four different heights and several different degrees of suspension hardness are available in order to adjust the device to the child's individual needs. Thanks to the different colours and fabric patterns the product can also be customised to the user's taste.

Produkt Product	Gentleman, 2008 Rollator / Walking frame
Hersteller Manufacturer	Meyra-Ortopedia, Kalletal-Kalldorf
Werksdesign In-house design	Andreas Nahrwold
Website Hersteller Manufacturer	www.meyra-ortopedia.de

Gentleman

Der Leichtbau-Rollator stellt sich dank seiner klassisch-eleganten Anmutung als unauffälliges Hilfsmittel dar. Durch seine zurückgenommene Erscheinung tritt er im Gebrauch in den Hintergrund, um die Aufmerksamkeit auf seinem Anwender zu belassen. Seine aufrechte und selbstbewusste Gestaltung weist unmittelbar auf sein Einsatzgebiet hin, während die integrierten Funktionen farblich hervorgehoben sind, sodass sie sofort identifiziert werden können. Rein der Statik dienende Elemente sind reduziert gestaltet, vermitteln aber durch ihre geschwungenen Linien eine angenehme Handhabung. Die ergonomisch ausgeführten und anatomisch geformten Handgriffe in Verbindung mit den benutzerfreundlichen Bremsen versprechen Sicherheit im Alltag. Die Verwendung wertiger Materialien wie glasfaserverstärkter Kunststoffe und hochfester Aluminiumlegierungen trägt zur Stabilität des Produktes bei und ermöglicht sein geringes Gewicht. Gleichzeitig unterstreichen sie seine Robustheit und Langlebigkeit. Der Rollator lässt sich aufgrund der höhenverstellbaren Schiebegriffe an die Körpergröße seines Benutzers anpassen. Durch sein kleines Packmaß im zusammengefalteten Zustand ist er leicht zu transportieren und zu verstauen.

Its classic, elegant look makes the lightweight walking frame a discreet aid. Thanks to its unobtrusive appearance it blends into the background, so that the attention remains on the user. Its upright and self-confident form clearly indicates its purpose and the integrated functions are emphasised with the help of colours. Elements which have a purely structural function have been given a streamlined form, nevertheless conveying a pleasant use with its curved lines. Ergonomically designed and anatomically shaped handlebars in combination with user-friendly brakes ensure safety in everyday use. The consistent use of high-quality materials such as glass fibre reinforced plastic and high-strength aluminium alloys contributes to the product's stability and light weight. At the same time they emphasise its robustness and durability. The walking frame can be adjusted to the user's height via the adjustable sliding handles. Thanks to its small size when folded together it is easy to transport or store.

Produkt Product	Swing, 2006 Treppenlift / Stair lift
Hersteller Manufacturer	ThyssenKrupp Accessibility BV, Krimpen aan den IJssel, Niederlande / The Netherlands
Design	MMID full service design team BV, Delft, Niederlande / The Netherlands
Website Hersteller Manufacturer	www.tkacc.nl
Website design	www.mmid.nl

Swing

Bei Hilfsmitteln, die als fester Bestandteil in das Wohnumfeld ihres Nutzers integriert sind, kommt der Gestaltung eine besondere Bedeutung zu. Dank seiner skulpturalen Formensprache, die mittels eines innovativen Spritzgießverfahrens realisiert werden konnte, setzt der Treppenlift über seine Funktionalität heraus Akzente. Zudem kann er sich aufgrund seines auswechselbaren Bezuges an unterschiedliche Interieurs anpassen. Seine formale Selbstverständlichkeit verdeutlicht seine Funktion in einer ansprechenden Art. Zudem hebt die Gestaltung diejenigen Bereiche, die in Kontakt zum Anwender stehen, auf eine sympathische Weise hervor. Die technischen Merkmale des Produktes orientieren sich an den Bedürfnissen der Nutzer. Beispielhaft sind hier die Armlehnen zu nennen, die sich als Sitz- und Aufstehhilfe verwenden lassen und gleichzeitig durch ihre gebogene Gestaltung ein Gefühl der Sicherheit vermitteln. Weiterhin tragen der Einklapphandgriff und der abnehmbare Bezug dem ergonomischen Aspekt Rechnung. Dank seines kompakten Antriebsgehäuses und des kleinen Wendekreises ist der Lift auch für beengte räumliche Verhältnisse geeignet und lässt sich in den unterschiedlichsten Treppenhäusern montieren.

The design of aids which are integrated into the user's living environment as permanent features has special significance. Thanks to its sculptural use of forms, which was created using an innovative injection moulding process, the stair lift sets accents beyond its functionality. Furthermore, with its removable cover it can adjust to different interiors. Its natural form presents its function in an appealing way. In addition, the design attractively highlights the areas that come into contact with the user. The product's technical features are geared towards meeting the users' needs; the armrests, for instance, can be used as support when sitting down or standing up, and with their bent shape they convey a sense of safety. Furthermore, the retractable handle and the removable cover take into account ergonomic aspects. Thanks to its compact drive housing and small turning circle, the lift is also suitable for confined spaces and can be installed in a great variety of staircases.

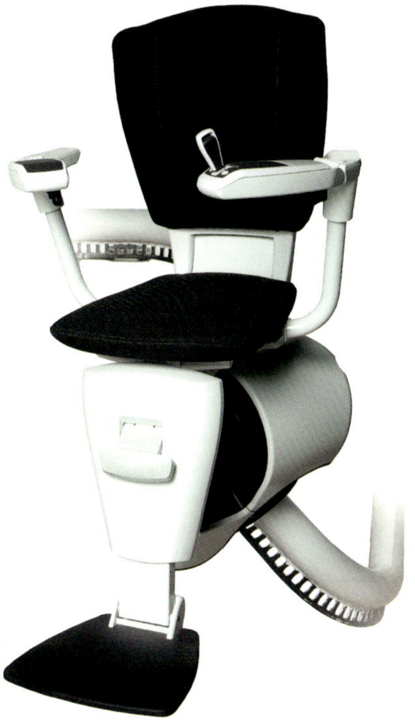

GOOD DESIGN
GEWINNER WINNERS

Produkt Product	RIFTON Sololift, 2007 Lifter
Hersteller Manufacturer	RIFTON Community Products LLC, Chester (NY), USA
Design	Werksdesign / In-house design
Vertrieb Distributor	Schuchmann GmbH & Co. KG, Hasbergen
Website Hersteller Manufacturer	www.rifton.com
Website Vertrieb Distributor	www.schuchmann-reha.de

RIFTON Sololift

Die Bewegung von Patienten ist für Betreuungspersonen nicht selten mit einer körperlichen Anstrengung verbunden. Auch kann der körperliche Kontakt zu fremden Personen vom Pflegebedürftigen als unangenehm empfunden werden. Als Alternative bietet dieser mobile Lifter die Möglichkeit, Patienten ohne Kraftaufwand aus oder in einen Stuhl oder Rollstuhl, ins Bett oder vom Boden zu heben. Seine zurückhaltende Farbgebung wirkt unaufdringlich, sodass der Nutzer im Vordergrund steht und nicht das Hilfsmittel. Einzig die gepolsterte Brustweste ist als das wichtigste funktionstragende Merkmal, das den Kontakt zum Körper herstellt, in einer auffälligeren Farbe gehalten. Die offene Konzeption des Lifters und die stabile Ausführung der tragenden Elemente vermitteln sowohl dem Patienten als auch dem Betreuer Vertrauen und Sicherheit. Der Lifter ermöglicht einen natürlichen Bewegungsablauf vom Sitzen zum Stehen und lässt sich ohne anzuecken durch normale Türen lenken. Bei Nichtbenutzung kann er dank seiner kompakten Gestaltung unauffällig untergebracht werden.

Moving patients often requires physical effort on behalf of caregivers. Furthermore, the physical contact with strangers may be experienced as unpleasant by patients. As an alternative, this mobile lifter offers the opportunity to lift patients effortlessly out of or into a chair or wheelchair, into a bed or up from the floor. Its discreet colour scheme makes it look unobtrusive, emphasising the person instead of the aid. As the most important functional element which makes contact with the body, the padded chest vest is the only component in a more conspicuous colour. The lifter's open design and the robust construction of the supporting elements instil a sense of trust and safety in patient and nursing staff alike. The lifter facilitates a natural movement from sitting to standing and can be easily steered through standard-sized doors. Its compact format allows it to be stored unobtrusively when not in use.

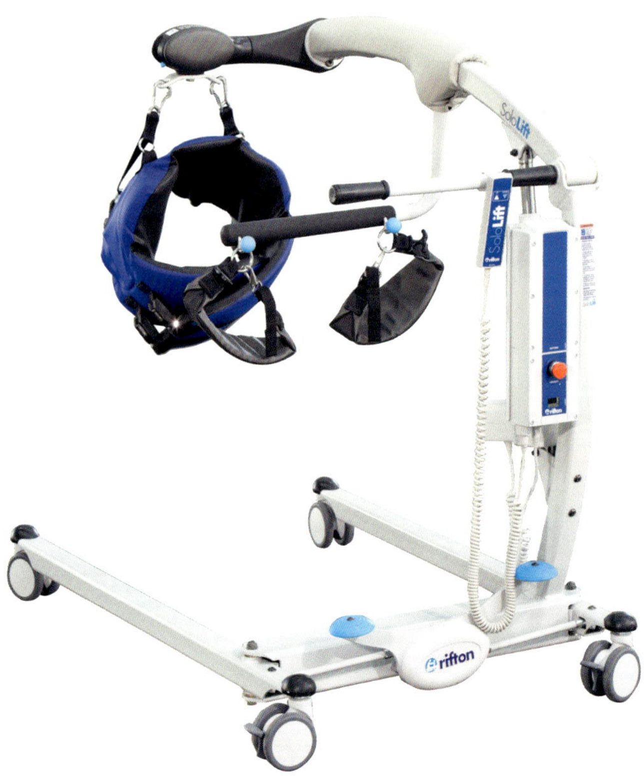

Produkt Product	BALANCE-Trainer, 2007 Dynamischer Steh- und Balance-Trainer Dynamic standing and balance trainer
Hersteller Manufacturer	medica Medizintechnik GmbH, Hochdorf
Werksdesign In-house design	Andreas Mack (Produktmanager), Otto Höbel (Technischer Leiter) Peter Kopf (Geschäftsführer), Klaus Weber (Konstruktion)
Website Hersteller Manufacturer	www.thera-trainer.de

BALANCE-Trainer

Dieser dynamische Stehtrainer vereinigt das therapeutisch starre Stehen mit einem aktiven Stehen. Seine technische Anmutung weist direkt auf seinen funktionalen Charakter hin und zeigt sie in einer auffälligen Weise: Alle technischen Details sind präsent, sodass sich die Möglichkeiten der Anwendung sofort erschließen. Entsprechend markiert seine in sich schlüssige Formensprache eine deutliche Klarheit. Seine formale Eigenständigkeit erreicht das Produkt über die Details, die in einem direkten Kontakt zum Benutzer stehen: Der Holztisch mit seinen angenehm abgerundeten Kanten wie die in Leder ausgeführten Führungselemente betonen die Schnittstelle zum Anwender auf eine sympathische Weise und schaffen dadurch eine optische Zutraulichkeit. Zum anderen haben die Führungselemente einen wichtigen funktionstragenden Charakter, da sie entscheidend den Benutzerkomfort beeinflussen. Die Abstimmung des Produktes auf unterschiedliche Bedürfnisse wurde so einfach wie möglich gehalten. Beispielgebend stehen hier die Sterngriffe zur Höheneinstellung oder die verschiebbaren Polster der Hüft- und Kniepelotten, deren Einstellung ohne Werkzeug vorgenommen werden kann.

This dynamic standing exerciser combines the therapeutically rigid standing with an active form of standing. Its technical look reflects its functional character and presents it in an eye-catching way: all technical details are present, so that the possibilities of use become apparent immediately. Thus its coherent design language is characterised by a high level of clarity. The product achieves its formal independence with the details that are in immediate contact with the user: the wooden table with its pleasantly rounded edges as well as the leather guide elements emphasise the interface with the user in an appealing way, thus visually instilling trust. Furthermore, the guide elements have an important functional character since they make a significant contribution to user comfort. Adjusting the product to different requirements has been made as easy as possible. The star knobs for adjusting the height and the different movable cushions for the hip and knee pads, which can be adjusted without tools, are clear examples of this.

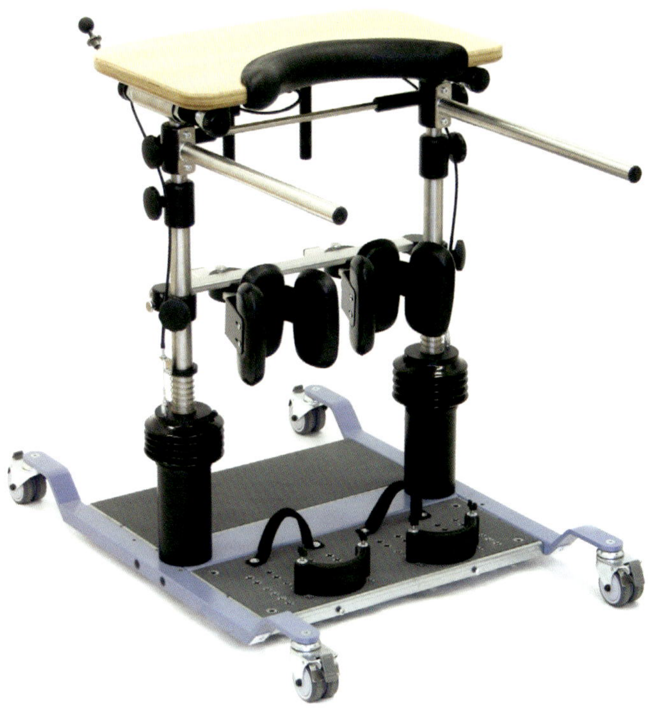

GOOD DESIGN
GEWINNER WINNERS

Produkt Product	TRlactive, 2006
	Orthopädische Einlage / Orthopedic insole
Hersteller Manufacturer	Bauerfeind AG, Zeulenroda-Triebes
Werksdesign In-house design	Ralf Lang
Website Hersteller Manufacturer	www.bauerfeind.com

TRlactive

Die Drei-Zonen-Einlage zeigt schon in der Optik ihre Funktionalität. Das Besondere: Die Einlage ist vertikal in farblich abgesetzten Zonen aufgebaut, welche verschiedenen Härtegraden entsprechen und in der einmaligen Puzzleoptik übergangslos zusammenwirken. Durch die Farbgebung erhält das Produkt eine für den Bereich der orthopädischen Einlagen ungewohnt frische Anmutung. Die Drei-Zonen-Einlage wurde für die Bedürfnisse aktiver Menschen entwickelt. Den beim Sport auftretenden Druck- und Torsionskräften wirkt das Einlagenkonzept effektiv entgegen. Die härteren Zonen stützen den Fuß, während die weicheren Zonen Druckspitzen gezielt verteilen und den Fuß damit entlasten. Für unterschiedliche Sportarten mit ihren spezifischen Bewegungsmustern wurden Einlagen in verschiedenen Ausführungen entwickelt. Damit wird den differenzierten Belastungen der Fußzonen, wie sie beispielsweise beim Laufen oder bei Ballsportarten auftreten, Rechnung getragen.

The appearance of the three-zone insole already demonstrates its functionality. Its special feature is its insole which consists of vertically colour-coded zones that correspond to different degrees of hardness and with their unique puzzle look work together seamlessly. The colour scheme brings an unusual freshness to the area of foot orthotics. The three-zone insole has been developed for the requirements of active people with the aim to support healthy running. The insole concept effectively counteracts the forces of pressure and torsion that occur during sport. The harder zones support the foot, while the softer zones provide targeted pressure distribution thus relieving the foot. A variety of insoles have been developed for different sports and their specific movement patterns. This takes into account the different kinds of strain on the different zones, as they occur in running or ball sports for instance.

Produkt Product	Thevo autoactiv, 2007 Anti-Dekubitus-System / Anti-decubitus aids
Hersteller Manufacturer	Thomas Hilfen für Körperbehinderte GmbH & Co. Medico KG, Bremervörde
Werksdesign In-house design	Dr. Klaus Jansen, Anneke Wilhelm
Website Hersteller Manufacturer	www.thomashilfen.de

Thevo autoactiv

Das Bewegungsbett wurde entwickelt, um Druckgeschwüre bei bettlägerigen Patienten zu verhindern und deren Behandlung zusätzlich zu unterstützen. Die differenzierte Gestaltung von Unterfederung und Flügelfedern verleiht dem Produkt eine luftige Anmutung, die den Einsatzzweck positiv unterstreicht. Die angenehm abgerundeten Flügelfedern nehmen die kleinsten Bewegungen des Patienten, wie Atembewegungen, Muskelanspannungen oder den Herzschlag, auf und verwandeln sie in feine Gegenbewegungen. Durch diese MiS Micro-Stimulation werden die Wahrnehmung des Nutzers stimuliert und seine Eigenbewegungen verbessert. Das sensible Lagerungssystem unterstützt die Dekubitusbehandlung, indem der Patient weich und bewegt zugleich liegt. Druckspitzen werden von der einheitlichen Auflagefläche vermieden, gleichzeitig wird die Restmobilität nicht eingeschränkt, sondern positiv stimuliert und gefördert. Mehr Bewegung minimiert die Dekubitusgefahr. Das Bewegungsbett passt in jedes vorhandene Bett.

The movement bed was developed in order to prevent pressure ulcers in bed-ridden patients and additionally support their treatment. The sophisticated design of the suspension system and the wing-shaped elements gives the product an airy look, which positively emphasises its use. The pleasantly rounded wing suspension elements absorb the smallest movements of the patient such as breathing, flexing of muscles, or heart beats, and turn them into fine counter movements. This MiS micro-stimulation stimulates the users' perception and increases their own movements. The sensitive suspension system supports the treatment of bedsores by providing a soft and moving surface for the patient. Pressure points are prevented by this uniform surface; at the same time the remaining mobility is not restricted but stimulated positively and supported. Increased movement minimises the risk of bedsores. The movement bed fits into any existing bed.

Produkt Product	Pibella Comfort, 2008 Urinierhilfe für Frauen / Female urination system
Hersteller Manufacturer	Stebler.net GmbH, Wettingen, Schweiz / Switzerland
Werksdesign In-house design	Rene und Fabian Stebler
Vertrieb Distributor	Foxy Ltd for Pibella, Hongkong
Website Hersteller Manufacturer	www.pibella.com

Pibella Comfort

Die Versorgung von pflegebedürftigen Menschen im Intimbereich zählt zu den sensibelsten Aufgaben für Pflegekräfte. Umso wichtiger ist es hier, Hilfsmittel zur Verfügung zu haben, die eine natürliche und diskrete Handhabung erlauben. Die Urinierhilfe bietet eine ästhetisch ansprechende Lösung, um Frauen das Wasserlassen im Liegen zu erleichtern. Ihre schlichte Form steht dabei in einem Gegensatz zu ihrem hohen Nutzen bei der Anwendung. Die organische Gestaltung offenbart ihre Funktionalität und ermöglicht eine intuitive Handhabung von den Patientinnen selbst sowie auch seitens des Pflegepersonals. Die aus einem hautfreundlichen Material bestehende Urinierhilfe wurde an die anatomischen Gegebenheiten von Frauen angepasst, sodass sie sich ohne Aufwand ansetzen und nutzen lässt. Das sich nach unten hin verjüngende Rohr aus Polypropylen wird dabei fest in einen Urinbeutel eingesetzt, der mit einem Deckel hygienisch verschlossen werden kann. Das Produkt ist ein gutes Beispiel für eine gestalterisch einfach scheinende Lösung, die einen vielfältigen Nutzen für seine Anwender und Betreuungspersonen bietet.

Providing genital care to people in need of care is one of the most sensitive tasks for nursing staff. Therefore it is all the more important to have aids available which allow natural and discreet use. This urination system offers an aesthetically appealing solution allowing women to pass urine when lying down. Its plain form stands in direct contrast to its high utility. The organic design reveals its functionality and enables intuitive use by patients and nursing staff alike. The urination device is made of a skin-friendly material and has been adjusted to the female anatomy, so that it can be positioned and used effortlessly. The downward tapered tube made of polypropylene is tightly inserted into the urine bag, which can be hygienically sealed with a lid. The product is a good example of what appears to be a simple solution in terms of design and which offers many advantages for its users and nursing staff.

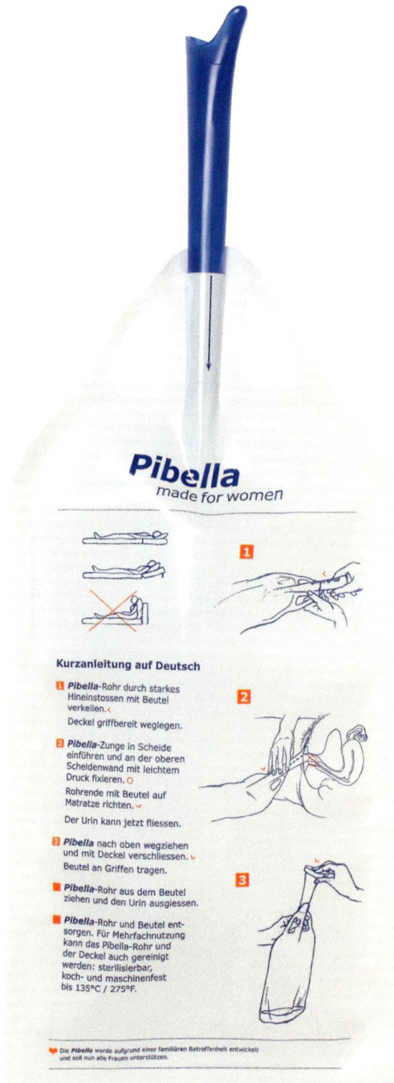

Produkt Product	Pibella Travel, 2008 Urinierhilfe für Frauen / Female urination system
Hersteller Manufacturer	Stebler.net GmbH, Wettingen, Schweiz / Switzerland
Werksdesign In-house design	Rene und Fabian Stebler
Vertrieb Distributor	Foxy Ltd for Pibella, Hongkong
Website Hersteller Manufacturer	www.pibella.com

Pibella Travel

Die Urinierhilfe für Frauen ist ein hochintegratives Produkt, da sie sich sowohl für behinderte Menschen eignet als auch in gewissen Situationen von jeder Frau verwendet werden kann. Es offenbart sich als eine praktische Lösung für den intimsten Bereich. Durch seine positive Formensprache und organische Gestaltung erweckt es Vertrauen bei seinen Nutzerinnen – und schafft dadurch eine Grundvoraussetzung für eine selbstverständliche Anwendung. Die Urinierhilfe bietet in Umgebungen, in denen der Hautkontakt mit Sanitäreinrichtungen unerwünscht sein kann, eine hygienische und diskrete Alternative. Auch bei beengten Verhältnissen ist ihr Einsatz möglich. Dank seiner einfachen Handhabung und sympathischen Anmutung kann das Produkt das spezifische Gebiet eines Reha-Hilfsmittels verlassen und in verschiedenen anderen Bereichen, beispielsweise bei Trekking-Touren, verwendet werden. Das Röhrchen aus hautfreundlichem Material wird im Sitzen oder Stehen angewendet. Die weiche Ausformung ist den anatomischen Gegebenheiten angepasst und verhindert, dass Urin an unerwünschten Stellen austritt.

This female urination system is a highly integrative product since it is suitable for disabled people and can be used by any woman in certain situations. It is a practical solution for the most intimate area. Its positive design language and organic form instil trust in its users, thus creating a basic prerequisite for natural use. In environments, in which skin contact with sanitary appliances can be problematic, the urination system offers a hygienic and discreet alternative. It can also be used in confined spaces. Thanks to its easy use and pleasant appearance the product is not restricted to the specific area of rehabilitation, but can be used in a variety of different situations such as, for example, on trekking tours. The tube made of skin-friendly material can be used in a standing or sitting position. The soft shape was designed with the female anatomy in mind and prevents unwanted urine spillage.

Produkt Product	DynaVox M³, 2008 Kommunikationshilfsmittel / Communication device
Hersteller Manufacturer	Dynavox Technologies, Pittsburgh, USA
Design	Insight Design, Maynard (MA), USA
Vertrieb Distributor	REHAVISTA GmbH, Berlin
Website Hersteller Manufacturer	www.dynavoxtech.com
Website Design	www.insightpd.com
Website Vertrieb Distributor	www.rehavista.de

DynaVox M³

Die elektronische Kommunikationshilfe zeigt eine sympathische Form und stellt dadurch eine ansprechende Verbindung zwischen dem technischen Instrument, seinem Nutzer und dem Gesprächspartner her. Ihre zeitgemäße Gestaltung mit den abgerundeten Kanten und die Farbgebung in modernem Weiß tragen zu einer höheren Akzeptanz seitens der vornehmlich jungen Nutzergruppe bei. Das auf Symbolbasis funktionierende Gerät gibt nach Druck auf das Display Wörter, Sätze oder komplexe Aussagen wieder. Durch seinen einfachen Aufbau ist es sowohl von Kindern als auch von Erwachsenen intuitiv zu bedienen und für die grundlegende Bedürfnisäußerung bis hin zu einer komplexen Kommunikation geeignet. Interaktive Szenenbilder visualisieren den Kontext der jeweiligen Kommunikationssituation, sodass sich Menschen mit Problemen im Sprachverständnis leichter in einer Gesprächssituation zurechtfinden. Zielgruppen sind Kinder, die die Schriftsprache nicht oder noch nicht beherrschen, Erwachsene mit eingeschränkter Sprechfähigkeit oder Menschen mit kognitiven Einschränkungen.

The electronic communication aid has an attractive form, thus creating an appealing connection between the technical device, its user and the conversation partner. Its contemporary design with rounded edges and the modern white colour increase its acceptance by the predominantly younger user group. The device, which works on the basis of symbols, plays words, sentences or complex statements at the touch of a button on the display. Thanks to its simple design it can be used intuitively by children and adults alike and is suitable for communication ranging from uttering basic needs to complex situations. Interactive visual scenes display the context of the respective communication situation, thus enabling people with speech comprehension problems to follow the conversation more easily. Target groups are children who cannot master or have not yet mastered written language, adults with a limited ability to speak and people with cognitive restrictions.

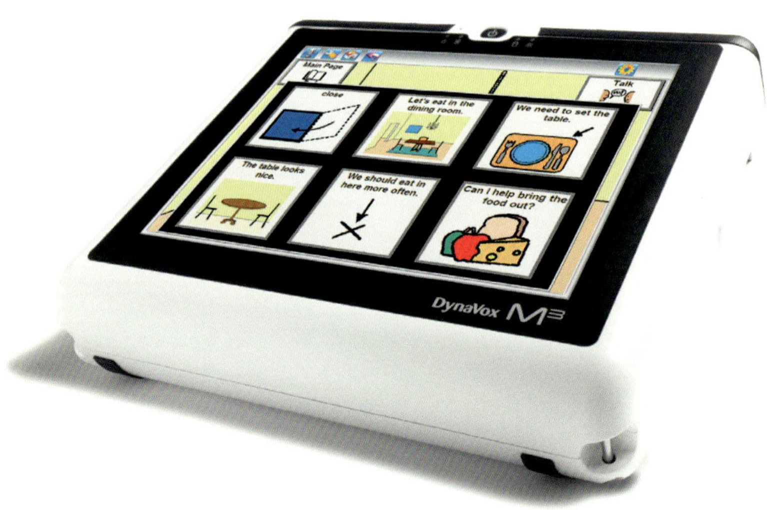

Produkt Product	TypeSpeak, 2007 Schriftbasiertes elektronisches Kommunikationshilfsmittel (Software) Electronic text-to-speech communication aid (software)
Hersteller Manufacturer	Phoenix Software GmbH, Bonn
Design	Werksdesign / In-house design
Website Hersteller Manufacturer	www.phoenixsoftware.de

TypeSpeak

Das schriftbasierte elektronische Kommunikationshilfsmittel ist ein Beispiel dafür, wie modernste Technologie in einer intelligenten und sympathischen Weise beherbergt werden kann. Es zeigt eine Handlichkeit, dank derer es sich in jeder Situation im Alltag einsetzen lässt. Das Gerät ist leicht und besitzt einen Klappmechanismus, sodass der Anwender es bequem transportieren oder in einer Tasche verstauen kann. Dadurch bietet es die Voraussetzungen, um ein ständiger Begleiter von Menschen zu sein, die in ihrer sprachlichen Kommunikationsfähigkeit eingeschränkt sind. Das Gerät zeigt eine technische Anmutung, wie sie auch moderne mobile Kommunikationsgeräte besitzen, und ist dadurch nicht unmittelbar als klassisches Hilfsmittel zu identifizieren, sondern wird zum hochwertigen Accessoire. Dank des Hilfsmittels haben nicht sprechende Menschen die Möglichkeit, ihre Mitteilungen über eine Tastatur einzugeben und vorlesen zu lassen. Die integrierte Software kann der Benutzer an seine Bedürfnisse anpassen und beispielsweise Schrift- und Tastengröße oder individuelles Vokabular einstellen. Um die Worteingabe zu erleichtern, verfügt die Software über eine Wort- und Buchstabenvorhersage sowie ein symbolunterstütztes Vokabular. Optionale Multimediafunktionen wie die SMS- oder Kamerafunktion erweitern die Bedeutung des Produktes über seine Eigenschaft als klassisches Hilfsmittel hinaus.

The type-based electronic communication aid is an example of how state-of-the-art technology can be integrated in an intelligent and appealing casing. Thanks to its handiness it can be used in any everyday situation. The device is light and has a folding mechanism, allowing the user to transport or store it away in a bag easily. Thus it fulfils all the requirements to be a constant companion for people who are limited in their ability to communicate through speech. The device has a technical look similar to that of modern mobile communication devices and is thus not immediately identifiable as a classic aid, but looks like a high-value accessory instead. Thanks to the aid people who do not speak are given the opportunity to enter their messages via a keyboard and have them read out. Users can adjust the integrated software to their needs and can, for instance, set font and key sizes or individual vocabulary. In order to make entering words easier the software has a predictive word/letter function as well as a symbol-based vocabulary. Optional multimedia functions such as SMS or camera functions increase the product's significance beyond its role as a classic aid.

REHACARE DESIGN AWARD
DIE JURY THE JURY

Garant für die Qualität des REHACARE DESIGN AWARD 2008/2009: Die Jury besteht aus international anerkannten Designexperten mit langjähriger Erfahrung im Medizin- und Rehabilitationsbereich.

Guarantor for the quality of the REHACARE DESIGN AWARD 2008/2009: The jury consists of internationally recognised design experts with many years of experience in the field of medical and rehabilitation products.

Francesco Milani

Mitbegründer der Milani design & consulting AG
Cofounder of the Milani design & consulting AG

Francesco Milani, Jahrgang 1937, besuchte die Kunstgewerbeschule in Basel und absolvierte von 1953 bis 1957 eine Ausbildung zum Grafiker mit einer anschließenden einjährigen Zusatzausbildung im Bereich Fotografie. Während seiner ersten vier Berufsjahre bei dem Architekten, Produktgestalter und Grafiker Franco Grigniani in Mailand entdeckte er seine Begeisterung für das Produktdesign. 1963 gründete er sein eigenes Büro für Industriedesign, Beratung und visuelle Kommunikation, wobei er sich vorrangig mit Hightech-Produkten, Medizintechnik, Elektrowerkzeugen für die Blechbearbeitung, Maschinenbau, Steuersystemen sowie Fotoblitzanlagen beschäftigte. Als Mitbegründer der 1997 ins Leben gerufenen Milani design & consulting AG erweiterte er bis 2002 sein Portfolio um die Marketing- und Kulturberatung sowie das Engineering bis zur Produktreife.

Francesco Milani, born in 1937, visited the Basel School of Design and trained as a graphic designer from 1953 to 1957, followed by a one-year internship in the field of photography. During his first four years of working with architect, product designer and commercial artist Franco Grigniani in Milan, he discovered his enthusiasm for product design. In 1963, he founded his own office for industrial design, consulting and visual communication services with a primary focus on high tech products, medical technology, electric power tools for sheet metal working, mechanical engineering, control systems, as well as photographic flashlight devices. As cofounder of Milani design & consulting AG, established in 1997, he extended his portfolio to include marketing and cultural consultation, as well as engineering up to product maturity until 2002.

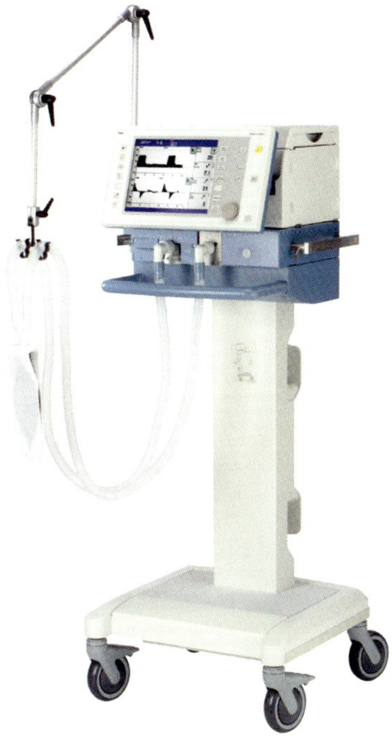

1

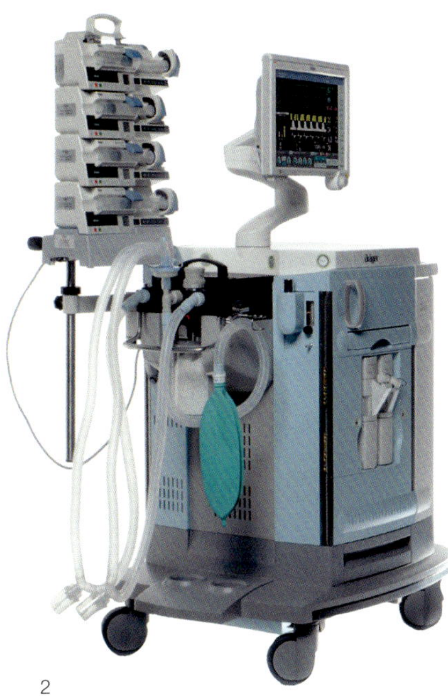

2

1 Evita® 4 edition, Intensiv-Beatmungsgerät / Evita® 4 edition, invasive ventilation machine
2 Zeus®, Anästhesiegerät / Zeus®, anaesthesia machine

Dirk Schumann

Gründer und Inhaber des Designstudios
Büro für industrielle Formentwicklung
Founder and owner of the design office
Büro für industrielle Formentwicklung

Dirk Schumann, 1960 in Soest geboren, studierte Produktdesign an der Fachhochschule Münster. Nach Abschluss seines Studiums im Jahr 1987 war er als Industriedesigner für OCO-DESIGN tätig und wechselte 1989 zu sieger-design. Bis 1991 war er an der Fachhochschule Münster als Lehrbeauftragter für Produktdesign tätig. Sein eigenes Unternehmen gründete Dirk Schumann 1992 in Münster unter dem Namen „Büro für industrielle Formentwicklung". Der Schwerpunkt seiner Arbeit liegt auf der Entwicklung von Produkten aus den Bereichen Kommunikation, Medizintechnik, Forschung, Wissenschaft, Sanitär, Wohnen, Gastronomie und Freizeit, die ästhetische, fertigungstechnische und wirtschaftliche Aspekte optimal vereinen. Der mehrfach ausgezeichnete Designer beschäftigt sich ebenfalls mit konzeptioneller Architektur und entwirft visionäre Wohn- und Lebenswelten.

Dirk Schumann, born in Soest in 1960, studied product design at Münster University of Applied Sciences. After finishing his studies in 1987, he worked first as an industrial designer at OCO-DESIGN, then in 1989 moved to sieger-design. Until 1991, he was a lecturer of product design at Münster University of Applied Sciences. In 1992, Dirk Schumann founded his own company in Münster under the name "Büro für industrielle Formentwicklung". His work puts special emphasis on the development of products in the communication, medical technology, research, science, sanitary, living, catering and recreation businesses, as well as products that optimally combine all aspects of aesthetics, manufacturing and economics. Honoured with several awards, Schumann also focuses on conceptual architecture, designing visionary living and housing environments.

1

2

1 Puris, Wasserarmatur für den öffentlichen Bereich / Puris, industrial bathroom fixtures
2 Kamerasystem für Elektronenmikroskopie / Camera system for electron microscopy

Markus Wild

Geschäftsführer der WILDDESIGN medical
GmbH & Co. KG und Leiter der WILDDESIGN-Gruppe
CEO and Managing Director of WILDDESIGN medical
GmbH & Co. KG and head of WILDDESIGN group

Markus Wild, Jahrgang 1961, studierte an der Folkwang-Universität in Essen Industrial Design und absolvierte mit Diplom im Jahr 1988. Bereits während seines Studiums sammelte er als Freiberufler Erfahrungen unter anderem in der Siemens-Designabteilung in München. 1990 gründete er das inhabergeführte Designbüro WILDDESIGN. Die Spezialisierung auf Medical Design erfolgte 1996. Seit 2006 hat Markus Wild Dependancen in Hongkong, Shanghai sowie in Gelsenkirchen unter dem Dach der WILDDESIGN-Gruppe gegründet. Sein Unternehmen arbeitet heute in einem weiten Feld der Produkt- und Markenentwicklung unter dem Leitsatz: „We translate insight to design."

Markus Wild, born in 1961, graduated from the Folkwang University in Essen with a degree in industrial design in 1988. While still a student, he gained experience as a freelancer working in the Munich design department of Siemens. In 1990, he founded his own company, WILDDESIGN, and in 1996 the focus of the company became medical design. Since 2006, Markus Wild has founded subsidiaries in Hong Kong, Shanghai and Gelsenkirchen under the umbrella of the WILDDESIGN group. Today, the company works in a wide range of product and brand development under the motto, "We translate insight to design."

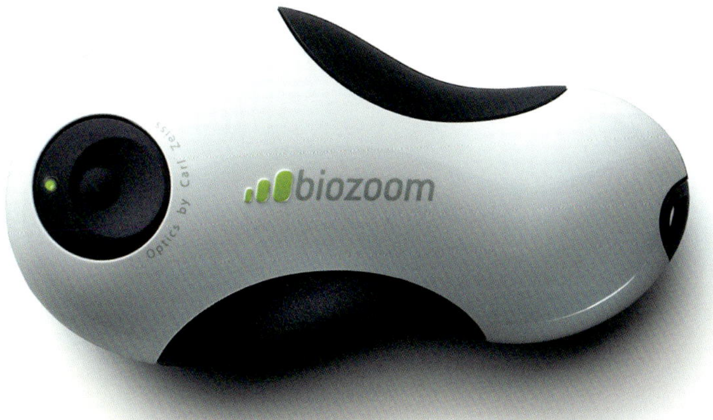

1

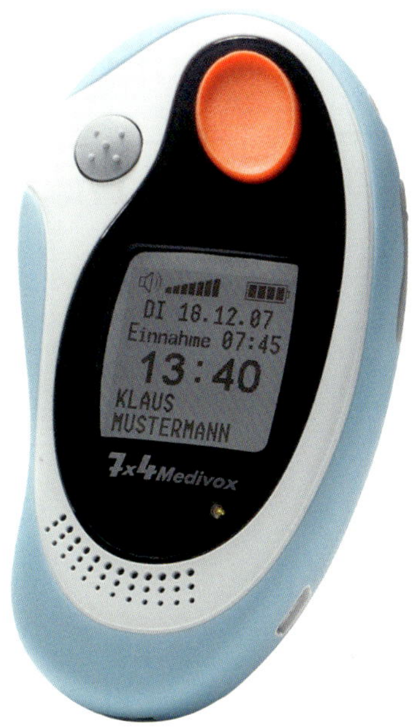

2

1 Biozoom, Handheld-Device zur nicht-invasiven Messung von Blutwerten / Biozoom, non-invasive, hand-held device for measuring blood values
2 Medivox, Sprachausgabegerät als Ergänzung zum 7x4 Wochenblister / Medivox, voice response device for use with the 7x4 blister pack

Prof. Dr. Peter Zec

Initiator des red dot design award und Senator des internationalen Dachverbandes der Industriedesigner Icsid (International Council of Societies of Industrial Design)
Initiator of the red dot design award and senator of the international umbrella organisation of industrial designers Icsid (International Council of Societies of Industrial Design)

Professor Dr. Peter Zec, geboren 1956, ist national und international einer der führenden Experten für Design und Kommunikation. Er studierte Medienwissenschaft, Psychologie und Kunstwissenschaft und wurde 1991 geschäftsführender Vorstand des international renommierten Design Zentrums Nordrhein-Westfalen. 1993 nahm er eine Professur für Wirtschaftskommunikation an der Hochschule für Technik und Wirtschaft in Berlin an. Seit 2001 ist er geschäftsführender Gesellschafter der red dot GmbH & Co. KG. Von September 2005 bis Oktober 2007 war er Präsident des internationalen Dachverbandes der Industriedesigner Icsid, von Oktober 2006 bis Oktober 2007 Vorsitzender der internationalen Designallianz IDA. Damit hatte er die zwei bedeutendsten Ämter der Designwelt inne. Heute hat er als Icsid-Senator maßgeblichen Einfluss auf die Entscheidungen des Verbandes. Seit mehr als 15 Jahren berät Peter Zec internationale Unternehmen in Fragen der Kommunikation und des Designs. Seine weltweite Anerkennung festigte er darüber hinaus in zahlreichen Vorträgen und durch den Aufbau eines weltumspannenden Netzwerks von Designern. Er ist Autor zahlreicher Bücher und Herausgeber des red dot design yearbooks sowie des international yearbook communication design.

Professor Dr. Peter Zec, born in 1956, is a leading national and international expert in design and communication. He studied media studies, psychology and art theory. In 1991, he became president of the internationally renowned Design Zentrum Nordrhein-Westfalen, and in 1993 accepted a professorship for economic communication at the University of Applied Sciences for Technology and Economics in Berlin. Since 2001, he has been managing partner of red dot GmbH & Co. KG. From September 2005 to October 2007, Zec was president of the International Council of Societies of Industrial Design (Icsid), and from October 2006 to October 2007 chairman of the International Design Alliance (IDA), thus holding two of the highest official posts in the design world. As Icsid Senator he is now consulted on important issues that concern the association. For more than 15 years, Peter Zec has acted as a consultant to international businesses, sharing his experiences on communication and design issues. His reputation further rests on numerous lectures given throughout the world and his leading role in the development of a global network of designers. He is author of numerous publications, as well as editor of the red dot design yearbook and the international yearbook communication design.

1

2

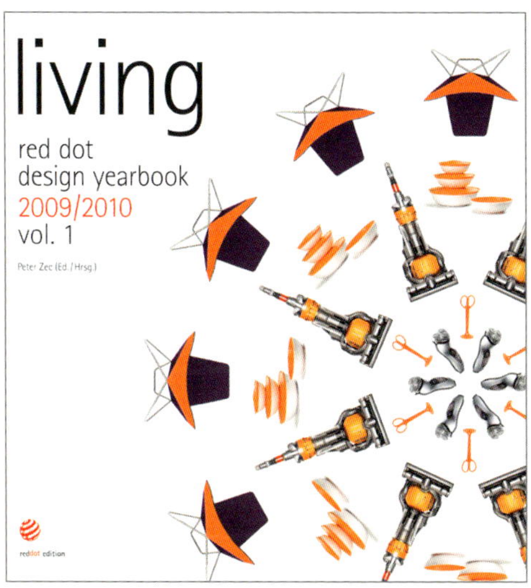

3

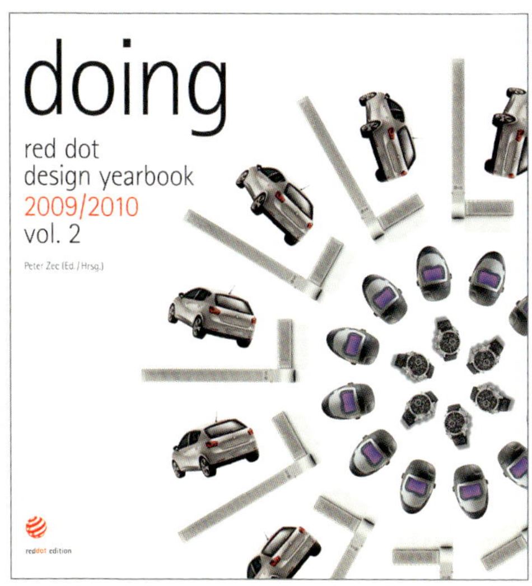

4

1 Who's Who in Design – The Leading Designers of the World. Volume 3
2 international yearbook communication design 2009/2010
3 red dot design yearbook 2009/2010: Volume 1: living
4 red dot design yearbook 2009/2010: Volume 2: doing

ANHANG / APPENDIX
QUELLEN SOURCES

Hinweise zu den Autoren
About the authors

Eckhard Feddersen

Eckhard Feddersen, geb. 1946, absolvierte sein Architekturstudium an den Technischen Universitäten in Karlsruhe und Berlin. 1973 gründete er gemeinsam mit Wolfgang von Herder ein eigenes Architekturbüro in Berlin. 2002 fokussierte er sich mit seinem neu gegründeten Architekturbüro Feddersen Architekten weiter im Bereich Sozialimmobilien. Neben diversen Fachartikeln und Buchbeiträgen veröffentlichte er 2009 gemeinsam mit Insa Lüdtke den „Entwurfsatlas Wohnen im Alter".

Eckhard Feddersen, born in 1946, completed his studies in architecture at the Technical Universities in Karlsruhe and Berlin. In 1973, he and Wolfgang von Herder founded their own architectural firm in Berlin. In 2002, he opened a new firm, Feddersen Architekten, to continue working on social and public housing projects. In addition to numerous articles and book contributions, in 2009 he and Insa Lüdtke published, "Living for the Elderly: A Design Manual".

Burkhard Jacob

Burkhard Jacob, geb. 1969, absolvierte seine Ausbildung zum Werbekaufmann bei der renommierten Werbeagentur Grey Düsseldorf und studierte Wirtschaftskommunikation in Berlin. Seit 1997 arbeitet er als strategischer Berater für das Design Zentrum Nordrhein-Westfalen und die red dot design promotion in Essen. 2009 rief er gemeinsam mit Prof. Dr. Peter Zec das red dot institute for advanced design studies ins Leben, eine Service- und Forschungseinrichtung, die designorientierte Branchen und Unternehmen analysiert.

Burkhard Jacob, born in 1969, completed his training in advertising at the renowned advertising agency, Grey Düsseldorf, and studied business communications in Berlin. Since 1997, he has worked as a strategic consultant for the Design Zentrum Nordrhein-Westfalen and red dot promotion in Essen. In 2009, together with Prof. Dr. Peter Zec, he started the red dot institute for advanced studies – a service and research institute for analyzing design-focused companies and branches of business.

Insa Lüdtke

Insa Lüdtke, geb. 1972, absolvierte ihr Architekturstudium an der Technischen Universität Darmstadt und arbeitet seit 2000 als freie Journalistin an der Schnittstelle von Architektur, Wohnen und Gesundheit. Seit 2002 verantwortet sie den Aufbau und die Leitung der Öffentlichkeitsarbeit von Feddersen Architekten. Gemeinsam mit Eckhard Feddersen gründete sie 2008 das Beratungsunternehmen Cocon Concept Feddersen Lüdtke Beratung. Im Rahmen dieser Zusammenarbeit veröffentlichten sie 2009 als Herausgeber den „Entwurfsatlas Wohnen im Alter".

Insa Lüdtke, born in 1972, studied architecture at the Technical University in Darmstadt and has worked as a freelance journalist since 2000 covering architecture, living and health. Since 2002, she has been responsible for developing and heading the public projects department at Feddersen Architekten. She and Eckhard Feddersen founded their own consulting firm in 2008, Cocon Concept Feddersen Lüdtke Beratung. As part of their collaboration, the two published, "Living for the Elderly: A Design Manual" in 2009.

Prof. Dr. Birger P. Priddat

Birger P. Priddat, geb. 1950, ist Ökonom und Philosoph. Seit August 2007 ist er Inhaber des Lehrstuhls für Politische Ökonomie im Studium fundamentale an der privaten Universität Witten/Herdecke. Von August 2007 bis Dezember 2008 war er Präsident der Universität. Als Autor hat er mehr als 30 Bücher zu ökonomischen, ethischen und moralischen Fragestellungen veröffentlicht.

Birger P. Priddat, born in 1950, is an economist and philosopher. Since August 2007, he has held the Chair of Political Economics at the Private University of Witten/Herdecke. From August 2007 until December 2008 he was President of the University. As an author, he has published more than 30 books on economics, ethics and morals.

Stephan Scheuer

Stephan Scheuer, geb. 1965, ist Geschäftsfeldleiter für Ergonomie & Gebrauchstauglichkeit und zuständig für Produktprüfungen bei der TÜV Rheinland Group. Er leitet den Normungsausschuss für die Ergonomie von elektronischen Anzeigen beim DIN und ist Mitarbeiter bei ISO. Dort arbeitet er bei der Standardisierung von Usability-Normen für Produkte der Informations- und Kommunikationstechnik und für Medizinprodukte mit.

Stephan Scheuer, born in 1965, is a business leader in ergonomics and usability and is responsible for product testing at the TÜV Rheinland Group. He heads the standardization program for the ergonomics of electrical notifications at DIN, the German Institute for Standardization. He also works at ISO, the International Standardization Organization, contributing to the standardization of usability norms for information and communication products, as well as for medical products.

Dr. Reiner Wichert

Dr. Reiner Wichert ist Sprecher der Fraunhofer Allianz Ambient Assisted Living und koordiniert das Zukunftsthema „Assisted Personal Health“ für die Fraunhofer Gesellschaft. Er promovierte 2004 an der Technischen Universität Darmstadt im Bereich Computergrafik. Seit Januar 2005 ist er Leiter der Abteilung Interaktive Multimedia-Appliances des Fraunhofer Institutes für Graphische Datenverarbeitung (IGD) in Darmstadt mit dem Fokus auf Ambient Intelligence und Ambient Assisted Living.

Dr. Reiner Wichert is head of the Fraunhofer Alliance Ambient Assisted Living and coordinates the Fraunhofer frontline theme "Assisted Personal Health". He received his PhD in computer science from the University of Darmstadt in 2004. Since January 2005, he has been the head of the Interactive Multimedia Appliances department at the Fraunhofer Institute for Computer Graphics (IGD) in Darmstadt, specialising in Ambient Intelligence and Ambient Assisted Living.

Quellenverzeichnis
Bibliography

1 Johann Behrens: Gesundheitsapostel und Professionen im System pflegerischer gesundheitsförderlicher Unterstützung. In: Hallesche Beiträge zu den Gesundheits- und Pflegewissenschaften. Tagungsband: Pflegebedürftig in der Gesundheitsgesellschaft. 6.-28. März 2009, S. 54
Seite / page 30

2 German Foundation for World Population, 2008
Seite / page 30

3 David M. Cutler: Das amerikanische Gesundheitssystem. In: Medical Solutions, Juli 2008, S. 20
Seite / page 31

4 Charles Taylor: Das Unbehagen an der Moderne. Frankfurt am Main 1995
Seite / page 31

5 David M. Cutler: Das amerikanische Gesundheitssystem. In: Medical Solutions, Juli 2008, S. 21
Seite / page 31

6 Volker Eric Amelung / Katharina Janus: Gesundheit vor der Wahl. In: Gesundheit und Gesellschaft, Ausgabe 1/2008
Seite / page 32

7 Barrack Obamas Rede zur Gesundheitsreform vor der American Medical Association in Chicago am 15. Juni 2009. Redemanuskript veröffentlicht in der New York Times.
Internet-Quelle: http://www.nytimes.com/2009/06/16/health/policy/16obama.html?_r=1
Seite / page 34

8 Otto von Bismarck: Gesammelte Werke. Friedrichsruher Ausgabe von 1924/1935, Band 9, S. 195
Seite / page 35

9 Shingo Shimada / Christian Tagsold: Alternde Gesellschaften im Vergleich. Solidarität und Pflege in Deutschland und Japan. Bielefeld 2006, S. 50f.
Seite 36 / page 35

10 Shingo Shimada / Christian Tagsold: Alternde Gesellschaften im Vergleich, a. a. O., S. 77
Seite / page 36

11 Shingo Shimada / Christian Tagsold: Alternde Gesellschaften im Vergleich, a. a. O., S. 83
Seite 38 / page 36

12 Shingo Shimada / Christian Tagsold: Alternde Gesellschaften im Vergleich, a. a. O., S. 103
Seite 40 / page 38

13 Seiritsu Ogura: Das japanische Gesundheitssystem. In: Medical Solutions, März 2009, S. 66f.
Seite 41 / page 40

14 Department of Economic and Social Affairs Population Division
Internet-Quelle: http://www.un.org/esa/population/
Seite / page 42

15 United Nations enable,
Rights and Dignity of Persons with Disabilities
Internet-Quelle: http://www.un.org/disabilities/
Seite / page 43

16 Bundesministerium der Justiz
Internet-Quelle: http://www.gesetze-im-internet.de/bgg/index.html
Seite / page 44

17 U.S. Department of Justice
[ADA Americans with Disabilities Act]
Internet-Quelle: http://www.ada.gov/pubs/ada.htm
Seite / page 46

18 Bundesministerium der Justiz
[Verordnung zur Schaffung barrierefreier Informationstechnik nach dem Behindertengleichstellungsgesetz (Barrierefreie Informationstechnik-Verordnung - BITV)]
Internet-Quelle: http://www.gesetze-im-internet.de/bitv/BJNR265400002.html
Seite 47 / page 46

19 Emiliani, P. L. / Stephanidis, C.: Universal access to ambient intelligence environments: Opportunities and challenges for people with disabilities. IBM Systems Journal 44 (Issue 3), 2005, S. 605-619
Seite / page 46

20 Norgall, Thomas: Fit und selbständig im Alter durch Technik – von der Vision zur Wirklichkeit? Bundesgesundheitsblatt Band 52, Nummer 3, Berlin 2009, S. 297-305
Seite / page 54

21 Heinze, Rolf G. et al.: Neue Wohnung auch im Alter – Folgerungen aus dem demographischen Wandel für Wohnungspolitik und Wohnungswirtschaft. Schader-Stiftung. Darmstadt 1997
Seite 55 / page 54

22 Robert Koch Institut: Themenheft 10: Gesundheit im Alter, Gesundheitsberichterstattung des Bundes. Berlin 2005a
Seite / page 55

23 Aarts, E. / Encarnação, J. L.: Into Ambient Intelligence
In: True Visions: Tales on the Realization of Ambient Intelligence.
E. Aarts and J. Encarnaçao (Hrsgg.).
Berlin / Heidelberg / New York 2005
Seite / page 58

24 Wichert, Reiner / Tazari, Mohammad-Reza / Hellenschmidt, Michael: Architectural Requirements of Ambient Intelligence.
In: it - Information Technology, 2008/1, S. 13-20
Seite / page 58

25 Wichert, R.: The Fraunhofer Alliance „Ambient Assisted Living" – Fraunhofer Allianz Ambient Assisted Living.
In: Microsystems Technology in Germany. 2008, pp. 26-27,
Editor Trias Consult, Berlin 2008
Seite 59 / page 58

26 BMBF-Studie Marktpotenziale, Entwicklungschancen, Gesellschaftliche, gesundheitliche und ökonomische Effekte der zukünftigen Nutzung von Ambient Assisted Living (AAL)-Technologien, Fraunhofer IGD / BIS / IGES
Seite 61 / page 60

Bildnachweis
Photo credits

Seite / page 4
Messe Düsseldorf GmbH, Düsseldorf:
Wilhelm Niedergöker

Seite / page 6, 126
red dot GmbH & Co. KG, Essen:
Peter Zec

Seite / page 18
Carl Zeiss Archiv, Jena:
Nachbildung eines Lesesteins / Replication of a reading stone

Science & Society Picture Library, London:
Stirnreifenbrille aus dem Jahr 1796 / Headband spectacles, 1796

Seite / page 19
Science & Society Picture Library, London:
Ohren- oder Schläfenbrille, 1800-1802
Transverse folding spectacles, 1800-1802

Nasenkneifer (Brille ohne Bügel) mit Etui, 1870-1920
Pince-nez spectacles with case, 1870-1920

Seite / page 20
Wikipedia:
„Eiserne Hand" des deutschen Ritters Götz von Berlichingen, 1504-1510 / „Iron Hand" by German knight Götz von Berlichingen, 1504-1510

Science & Society Picture Library, London:
Künstlicher Arm mit Haken und Handprothese, 1924-1930
Artificial arm with hook and forearm attachments, 1924-1930

Seite / page 21
Otto Bock HealthCare GmbH, Duderstadt:
C-Leg, das erste vollständig mikroprozessorgesteuerte Kniegelenk / C-Leg, the first fully microprocessor-controlled knee joint

DynamicArm, das weltweit erste elektronisch gesteuerte Elektro-Ellbogengelenk / DynamicArm, the first electronic-controlled elbow joint worldwide

Seite / page 44
Ströer Out-of-Home Media AG, Köln:
WC Tronic 402, vollautomatische barrierefreie Toilettenanlage
WC Tronic 402, fully automatic barrier-free sanitary system

Seite / page 47
Julius Blum GmbH, Höchst:
Servo-Drive, elektrische Bewegungs- und Öffnungsunterstützung
Servo Drive, electrical opening support system

Seite / page 50
Feddersen Architekten, Berlin:
Multifunktionale Fensterbank / Multifunctional windowsill

Orientierung in einem Etagenflur durch mehrere Raumsegmente
Several spatial segments provide orientation on each floor

Seite / page 65
Messe Düsseldorf GmbH, Düsseldorf:
REHACARE INTERNATIONAL 2008

Seite / page 70–75
Constantin Meyer, Köln:
REHACARE DESIGN AWARD 2008/2009

Seite / page 79–94
Werksfotos der Firmen / In-company photos:
Gewinnerprodukte und Designerporträts
Awarded products and designer portraits

Seite / page 98–117
Constantin Meyer, Köln:
Gewinnerprodukte / Awarded products

außer Seite / except page 98, 100, 102, 106, 109, 111
Werksfotos der Firmen / In-company photos:
Gewinnerprodukte / Awarded products

Seite / page 120, 122, 124
Constantin Meyer, Köln:
Jurorenporträts / Jury portraits

Seite / page 121
©Dräger Medical AG & Co. KG, Lübeck:
Evita® 4 edition, Intensiv-Beatmungsgerät /
Evita® 4 edition, invasive ventilation machine
Zeus®, Anästhesiegerät / Zeus®, anaesthesia machine

Alle Rechte vorbehalten. Kein Teil darf ohne vorherige schriftliche Genehmigung wiedergegeben, in einem Datensystem gespeichert oder in irgendeiner Form, weder elektronisch noch mechanisch, durch Fotokopie, Aufnahme oder durch andere Art übertragen werden. / All rights reserved. No part of this publication may be reproduced, stored in a retrieval system, or transmitted, in any form or by any means, electronic, mechanical, photocopying, recording or otherwise, without prior written permission of Dräger Medical AG & Co. KG.

Seite / page 123
Schumann Büro für industrielle Formentwicklung, Münster:
Puris, Wasserarmatur für den öffentlichen Bereich
Puris, industrial bathroom fixtures

Kamerasystem für Elektronenmikroskopie
Camera system for electron microscopy

Seite / page 125
WILDDESIGN medical, Gelsenkirchen:
Biozoom, Handheld-Device zur nichtinvasiven Messung von Blutwerten / Biozoom, non-invasive hand-held device for measuring blood values

Medivox, Sprachausgabegerät als Ergänzung zum 7x4 Wochenblister / Medivox, voice response device for use with the 7x4 blister pack

Hersteller- und Vertriebsregister
Manufacturer and distributor index

Ulrich Alber GmbH
Vor dem Weißen Stein 21
72461 Albstadt
Deutschland / Germany
www.alber.de
Seite / page 78–79

Bauerfeind AG
Triebeser Str. 16
07937 Zeulenroda-Triebes
Deutschland / Germany
www.bauerfeind.com
Seite / page 112

Coloplast GmbH
Kuehnstr. 75
22045 Hamburg
Deutschland / Germany
www.coloplast.de
Seite / page 92–93

Days Healthcare GmbH & Co. KG
Oberbecksener Str. 68
32547 Bad Oeynhausen
Deutschland / Germany
www.dayshealthcare.de
Seite / page 99

Dynavox Technologies
2100 Wharton Street, Suite 400
Pittsburgh, PA 15203
USA
www.dynavoxtech.com
Seite / page 116

Foxy Ltd for Pibella
112A Sheung Sze Wan Clearwaterbay,
NT Hong Kong
Hong Kong
www.pibella.com
Seite / page 114–115

Funwheelz
Kerkeind 7
5076 AA Haaren
Niederlande / The Netherlands
www.funwheelz.nl
Seite / page 102

Invacare Aquatec GmbH
Alemannenstr. 10
88316 Isny
Deutschland / Germany
www.invacare-aquatec.com
Seite / page 84–85, 98

Krabat AS
Ravnsborgveien 56
1395 Hvalstad
Norwegen / Norway
www.krabat.com
Seite / page 88–89, 107

Küschall AG
Benkenstr. 260
4108 Witterswil
Schweiz / Switzerland
www.kueschall.ch
Seite / page 84–85

medica Medizintechnik GmbH
Blumenweg 8
88454 Hochdorf
Deutschland / Germany
www.thera-trainer.de
Seite / page 111

Meyra-Ortopedia Vertriebsgesellschaft mbH
Meyra-Ring 2
32689 Kalletal/Kalldorf
Deutschland / Germany
www.meyra-ortopedia.de
Seite / page 108

Otto Bock Mobility Solutions GmbH
Lindenstr. 13
07426 Königsee
Deutschland / Germany
www.ottobock.de
Seite / page 100 – 101

Phoenix Software GmbH
Adolf-Hombitzer-Str. 12
53227 Bonn
Deutschland / Germany
www.phoenixsoftware.de
Seite / page 117

REHAVISTA GmbH
Pulsstr. 7
14059 Berlin
Deutschland / Germany
www.rehavista.de
Seite / page 116

RIFTON Community Products LLC
359 Gibson Hill Road Chester
NY 10918-2321
USA
www.rifton.com
Seite / page 110

Schuchmann GmbH & Co. KG
Dütestr. 3
49205 Hasbergen
Deutschland / Germany
www.schuchmann-reha.de
Seite / page 110

Stebler.net GmbH
Heimentalstr. 53
5430 Wettingen
Schweiz / Switzerland
www.pibella.com
Seite / page 114 – 115

Thomas Hilfen für Körperbehinderte GmbH & Co. Medico KG
Walkmühlenstr. 1
27432 Bremervörde
Deutschland / Germany
www.thomashilfen.de
Seite / page 103, 106, 113

ThyssenKrupp Accessibility BV
Van Utrechtweg 99
2921 LN Krimpen aan den IJssel
Niederlande / The Netherlands
www.tkacc.nl
Seite / page 109

Veigel GmbH + Co. KG
Lindenstr. 9-11
74653 Künzelsau
Deutschland / Germany
www.veigel-automotive.de
Seite / page 104 – 105

Designerregister
Designer index

einmaleins Büro für Gestaltung
Im Weitblick 1
88483 Burgrieden
Deutschland / Germany
www.einmaleins.net
Seite / page 78

Formel Industridesign AS
Sandakerveien 24C (Entry D7)
0473 Oslo
Norwegen / Norway
www.formel.no
Seite / page 88, 107

Insight Design
5 Clock Tower Place, Suite 120
Maynard, MA 01754
USA
www.insightpd.com
Seite / page 116

MMID full service design team BV
Westvest 145
2611 AZ Delft
Niederlande / The Netherlands
www.mmid.nl
Seite / page 109

RECARO GmbH & Co. KG
Competence Center Child Safety
Guttenbergstr. 14
95352 Marktleugast
Deutschland / Germany
www.recaro.com
Seite / page 103

Thamm Design
Hauptstr. 48
74595 Langenburg
Deutschland / Germany
www.thamm-design.de
Seite / page 104–105

IMPRESSUM IMPRINT

Herausgeber / Editor
Peter Zec

Projektleitung / Project supervising
Elmar Schüller

Projektmanagement / Project management
Dijana Milentijević

Projektleitung Messe Düsseldorf GmbH
Project management Messe Düsseldorf GmbH
Horst Giesen

Redaktion / Editorial work
Martina Stein, Otterberg

Lektorat / Proofreading
Ursula Heister, Köln

Übersetzung / Translations
Jan Stachel-Williamson, Christchurch, Neuseeland / New Zealand
Alexis O'Connolly, Essen
Andreas Zantop, Berlin
Christiane Zschunke, Frankfurt

Gestaltung / Layout
Markwald & Neusitzer Kommunikationsdesign, Frankfurt

Produktion, Lithografie und Druck
Production, lithography and printing
tarcom GmbH, Gelsenkirchen

reddot edition

Verlag / Publisher
red dot edition
Gelsenkirchener Straße 181
45309 Essen
Deutschland / Germany
T +49 201 8141-822
F +49 201 8141-810
reddot@red-dot.de
www.red-dot.de

ISBN 978-3-89939-112-1

avedition

Vertrieb weltweit / Worldwide distribution
avedition GmbH
Königsallee 57
71638 Ludwigsburg
Deutschland / Germany
T +49 7141 1477-391
F +49 7141 1477-399
kontakt@avedition.de
www.avedition.de

© 2009 red dot edition, Essen
Printed in Germany

Alle Rechte vorbehalten. / All rights reserved.

Stand bei Druck. – Alle Angaben ohne Gewähr.
Errors and omissions excepted. No liability can be accepted for any changes after the printing date.

Bibliografische Information der Deutschen Nationalbibliothek. Die Deutsche Nationalbibliothek verzeichnet diese Publikation in der Deutschen Nationalbibliografie; detaillierte bibliografische Daten sind im Internet über http://dnb.d-nb.de abrufbar.
Bibliographic information published by the Deutsche Nationalbibliothek. The Deutsche Nationalbibliothek lists this publication in the Deutsche Nationalbibliografie; detailed bibliographic data are available on the Internet at http://dnb.d-nb.de.